Nikola Tesla

The Life of a Genius and the Impact of His Work

(A Captivating Guide to the War of the Currents and the Life of Nikola Tesla)

James English

Published By **Frank Joseph**

James English

All Rights Reserved

Nikola Tesla: The Life of a Genius and the Impact of His Work (A Captivating Guide to the War of the Currents and the Life of Nikola Tesla)

ISBN 978-0-9958610-6-0

No part of this guidebook shall be reproduced in any form without permission in writing from the publisher except in the case of brief quotations embodied in critical articles or reviews.

Legal & Disclaimer

The information contained in this book is not designed to replace or take the place of any form of medicine or professional medical advice. The information in this book has been provided for educational & entertainment purposes only.

The information contained in this book has been compiled from sources deemed reliable, and it is accurate to the best of the Author's knowledge; however, the Author cannot guarantee its accuracy and validity and cannot be held liable for any errors or omissions. Changes are periodically made to this book. You must consult your doctor or get professional medical advice before using any of the suggested remedies, techniques, or information in this book.

Upon using the information contained in this book, you agree to hold harmless the Author from and against any damages, costs, and expenses, including any legal fees potentially resulting from the application of any of the information provided by this guide. This disclaimer applies to any damages or injury caused by the use and application, whether directly or indirectly, of any advice or information presented, whether for breach of contract, tort, negligence, personal injury, criminal intent, or under any other cause of action.

You agree to accept all risks of using the information presented inside this book. You need to consult a professional medical practitioner in order to ensure you are both able and healthy enough to participate in this program.

Table Of Contents

Chapter 1: Early Life And Education 1

Chapter 2: Early Engineering Career 16

Chapter 3: First Innovations And Ideas .. 26

Chapter 4: George Westinghouse 39

Chapter 5: Inventions And Discoveries... 56

Chapter 6: Colorado Springs Laboratory 73

Chapter 7: Financial Struggles And The Morgan Relationship.............................. 88

Chapter 8: The Grasp Of Lightning 104

Chapter 9: Death And The Aftermath .. 115

Chapter 10: Impact On Modern Science And Technology 124

Chapter 11: The Genius Innovator Ahead Of His Time 133

Chapter 12: What Tesla Gave The World? 148

Chapter 13: A Closer Look At Mr. Tesla!162

Chapter 14: The One Who Lit The World 178

Chapter 1: Early Life And Education
Childhood in Smiljan (1856-1873)

The tale of Nikola Tesla starts on July 10, 1856, within the quiet village of Smiljan, that's located in the Croatian area of the Austrian Empire. Tesla emerges as born amid an environment of rural simplicity, very unique from the technical marvels he could later construct. Milutin Tesla, his father, grow to be an eminent Orthodox priest famend for his scholarship and eloquence. Despite her loss of literacy, Đuka Mandić modified into a progressive housewife who made tool for the residence and had an innate ability to research Serbian epic poetry by manner of coronary coronary heart. Tesla maximum truly acquired his inventiveness and strong imagination from her.

Traditional values blended with highbrow stimulation to create the Tesla circle of relatives. Being the fourth child in a family of five, Nikola modified into raised in a domestic that placed a high significance on schooling and faith. He had a strict religious schooling sooner or later of his early years in Smiljan, and he have become curious approximately the herbal worldwide. Tesla's early years have been surrounded with the useful useful resource of the high priced, forested, and flow-crammed landscapes of the Lika area, which stoked his interest inside the workings of nature.

Tesla exhibited an super thoughts from an early age. He modified into able to carry out important calculus in his head, which led his instructors to really accept as true with he changed into cheating. His competencies in mathematics and physics had been apparent, and he showed a particular hobby in strength - a fascination that might define his lifestyles's artwork.

Despite the idyllic surroundings, Tesla's adolescence was not without tragedy. At the age of 7, he witnessed the loss of life of his older brother Dane in a horse-the use of coincidence. This disturbing event had a profound effect on Tesla, inducing visions and a heightened experience of sensitivity, factors that could play a enormous position in each his character and his technique to invention and discovery.

Nikola Tesla's Parent

The have an impact on of his mother and father changed into instrumental in shaping Tesla's adolescents. His father's splendid library furnished him with a wealth of information in philosophy, literature, and technological understanding, laying the muse for his lifelong love of studying. His mom's ingenuity and resourcefulness in fixing regular troubles with home made improvements undoubtedly stimulated Tesla's revolutionary trouble-fixing capabilities.

The Tesla domestic located a immoderate price on education. Seeing his son's amazing gadgets, Milutin Tesla preferred Nikola to grow to be a priest. But Tesla's hobby lay in the sciences. He begged his father to permit him pursue engineering faculty due to a scientific scenario he had as a teen, which also can had been cholera. His father gave in to desperation and agreed, setting Tesla on the short song to turn out to be one of the most progressive minds of all time.

During his childhood in Smiljan, Tesla modified into moreover uncovered to some of languages. The multicultural environment of the Austro-Hungarian Empire supposed that he have turn out to be fluent in severa languages, collectively with Czech, English, French, German, Hungarian, Italian, and Latin, in addition to his local Serbo-Croatian. This linguistic potential may additionally display worthwhile in his later travels and professional interactions in the course of Europe and America.

Nikola Tesla as a infant

Tesla's early experiments in Smiljan, even though rudimentary, were indicative of his later groundbreaking artwork. He have become interested by the phenomena of thunderstorms and the invisible forces of energy and magnetism. He might regularly be located experimenting with homemade electric powered gadgets, a good buy to the bewilderment of his circle of relatives and buddies.

As Tesla approached the stop of his time in Smiljan, it grow to be smooth that his destiny lay an extended manner past the pastoral hills of his location of starting place. With a thoughts brimming with ideas and a relentless hobby about the herbal international, he grow to be poised to step into the broader degree of clinical discovery and innovation. His next segment of schooling would possibly take him to the Austrian Polytechnic in Graz, a journey that would mark the beginning of his transformation from a curious boy from Smiljan to a pioneering inventor and electric powered engineer.

Thus, Tesla's great adventure began out with the early years in Smiljan. One day, amid the peace and tranquility of the Croatian geographical location, a small teenager with an insatiable hobby dreamed of controlling the forces of nature. The worldwide need to ultimately be revolutionized by using this dream.

Education and Early Influences

The story of Nikola Tesla places super emphasis on his training and early influences, which shed moderate on the makings of the mind that would move on to convert technological know-how and generation. Little Nikola have become the fourth of 5 kids born on July 10, 1856, within the tiny Austrian Empire city of Smiljan. His mom, Đuka Tesla, have become illiterate but had an extremely good reminiscence and a knowledge for building mechanical domestic domestic system. His father, Milutin Tesla, have become an Eastern Orthodox priest. Tesla commenced out his early schooling in this unassuming however intellectually appealing setting.

Tesla's early highbrow improvement have grow to be greatly aided via his mother. Her facile memorization of Serbian epic texts and her innovative use of household objects sincerely made an affect on younger Nikola. Đuka in no way received a right training, however her creativity and intellectual sharpness had a great effect on her son, training him to charge practical know-how and to rate the capability of the human thoughts.

Tesla's respectable schooling started out out in Smiljan, in which he attended fundamental faculty and showed early flair in technological understanding and mathematics. He delicate his academic capabilities at the adjacent Gospić higher faculty, in which he continued his schooling. However, Tesla regularly remembered with fondness how a bargain of a electricity Mark Twain's writings had on him at some point of these adolescence. In addition to offering a damage from his demanding instructional studies, Twain's writings inspired his creativity and sowed the seeds of mind that could ultimately blossom into his severa improvements.

Nikola Tesla's secondary college

In 1870, Tesla moved to Karlovac to attend the Higher Real Gymnasium, wherein he delved deeper into topics like physics and mathematics. Here, Tesla examined notable educational prowess, finishing a 4-one year software in most effective three years. It have become within the path of this period that Tesla first encountered the concept of strength - a phenomenon that could become the huge reputation of his life's artwork. The demonstrations of electrical experiments in his physics instructions captivated him, setting him on a route of inquiry and invention.

Following his graduation in 1873, Tesla again to Smiljan, in which he shrunk cholera. His infection emerge as immoderate and prolonged, and throughout this time, Tesla reportedly made a vow to his father that if he survived, he should grow to be an engineer - a promise that would steer the course of his destiny. After his healing, in 1875, Tesla enrolled at the Austrian Polytechnic in Graz, Austria, on a military frontier scholarship. At the Polytechnic, Tesla emerge as brought to using alternating current (AC) - an area in which he must later make huge contributions.

Tesla studied a superb deal eventually of his time on the Austrian Polytechnic. He changed into referred to for operating a strict agenda that proven his outstanding capacity for intellectual difficult paintings further to his love of mastering, lasting from three AM to 11 PM. Nevertheless, he had an entire lot lots less fulfillment in his second 12 months on the Polytechnic. Tesla lost his scholarship, have come to be addicted to gambling, and in the long run dropped out of faculty without incomes his diploma.

After leaving Graz, Tesla attended the Charles-Ferdinand University in Prague, even though he arrived too past due to join up and in no manner attended lectures regularly. Despite this, Tesla persisted his self-schooling, a testament to his insatiable thirst for understanding. He observe drastically, specializing in works related to power and engineering, laying the muse for his future enhancements.

The people and evaluations Tesla encountered at some point of his formative faculty years had a big impact at the improvement of the person who might probable later question ordinary medical information. His huge knowledge of electrical engineering and his progressive imagination, which drew idea from authors which encompass Mark Twain, laid the groundwork for a career that might be distinguished via way of both genius and controversy. It have grow to be Tesla who famously remarked, "The destiny, for which I surely worked, is mine; the winning is theirs." These traces perfectly capture the path of a young guy whose upbringing served as a basis for a legacy that might alternate the path of statistics.

University of Graz and Polytechnic School in Prague

University of Graz

During his time on the Polytechnic School in Prague and the University of Graz, Nikola Tesla's course into the fields of technological information and invention—which can ultimately make his call really etched in records—took a dramatic turn. These early years have been marked no longer fantastic thru academic hobbies however additionally through way of sizable intellectual and personal improvement, which laid the muse for his later ground-breaking discoveries.

Tesla's enrollment on the University of Graz in 1875 marked the start of his formal training in electric engineering. Here, in the hallowed halls and amidst the fervent intellectual ecosystem, Tesla grow to be first exposed to the dynamo, a device that converts mechanical electricity into electric powered energy. This publicity have end up pivotal, igniting in him a deep interest inside the capability of electricity, a region although in its infancy and riddled with mysteries and untapped opportunities. Tesla became an brilliant scholar, identified for his voracious appetite for knowledge and his near-obsessive have a take a look at conduct. He should frequently be placed poring over his textbooks and appealing in experiments, his mind continually at art work, grappling with the necessities of energy and magnetism.

However, Tesla's time at Graz have come to be not with out its disturbing situations. Financial troubles plagued his family, and Tesla, deeply aware of the sacrifices being made for his schooling, felt an outstanding pressure to excel. Despite those hardships, he remained undeterred, his remedy simplest strengthening his dedication to his studies.

In 1878, Tesla decided that could significantly impact his academic route. He left Graz and, for a time, severed formal ties with educational establishments. This duration changed into marked via private introspection and independent test, a testomony to Tesla's unyielding strength of mind to studying and discovery. He dove deeply into the test of energy and magnetism, exploring the works of scientists and philosophers who had laid the idea in these fields.

Tesla's academic adventure resumed at the Polytechnic School in Prague, albeit in short. His time there, although short-lived due to familial conditions, have become no matter the fact that impactful. It have grow to be proper proper right here that Tesla began to increase the foundational thoughts that might later become essential to his theories and upgrades. The rigorous curriculum and publicity to new clinical mind and theories supplied Tesla with a broader mind-set and a deeper facts of the capacity of electrical engineering.

These years at Graz and Prague have been greater than clearly durations of academic reading; they have been the crucible in which Tesla's specific method to technological expertise and invention grow to be stable. His reviews at these institutions did now not certainly equip him with the technical records and competencies that might be important in his later paintings, but similarly they formed his technique to hassle-solving and his philosophy closer to innovation. Tesla's time at the University of Graz and the Polytechnic School in Prague instilled in him a experience of wonder and hobby approximately the natural international, and a perception inside the trans formative power of energy and era.

The seeds of Tesla's destiny accomplishments had been planted during those early years. Tesla's path to turning into one of the maximum high-quality inventors of all time end up paved thru his immoderate instructional steering, character experiments, and inner pressure. His time at those universities had a wonderful feature in his growth as a scientist and an inventor, giving him the abilities and statistics he might want to convert electric powered powered engineering and alter the course of human history.

Chapter 2: Early Engineering Career

The preliminary years of Nikola Tesla's technical career, while his brilliance turn out to be nevertheless developing, had been spent deep in Europe. After completing his schooling, wherein his talents and nonconformist wondering had already began out to expose, Tesla started out a career that would sooner or later purpose his groundbreaking improvements.

In 1881, Tesla made his first foray into the engineering area outdoor of academia, first in Budapest, Hungary. His process at the Central

Telephone Exchange became proper right here. Around this time, his initial necessities regarding alternating present day (AC) began to tackle a extra tangible shape. Despite being little seemed, Tesla's work in Budapest become vital to growing his electric powered engineering competencies. It come to be additionally right here that he's said to have had the epiphany that introduced approximately the eventual revolution in the realm of energy: the idea of producing AC strength through a revolving magnetic location.

Image of even as Tesla lived in Paris

Image via Tesla Universe

In 1882, Tesla moved to Paris to paintings for the Continental Edison Company. This enjoy grow to be instrumental in exposing him to the burgeoning situation of electrical lighting fixtures and strength systems. While his tenure at Continental Edison end up a stepping stone in his career, it grow to be also a time of excessive mastering and announcement. Tesla's work normally involved installing indoor incandescent lighting systems, which have been all of sudden changing gas lighting in rich neighborhoods. This duration turn out to be marked via difficult paintings and prolific output, with Tesla regularly running from sunrise till a long way into the night time.

During this era, Tesla made contributions beyond clean implementation. His supervisors had been aware of his capacity to resolve difficult issues and enhance contemporary structures. One noteworthy example of his profound comprehension of electromagnetic necessities have become his remodel of inefficient dynamos. Tesla flourished in this ecosystem no matter the stressful difficult work and lengthy hours, and his enthusiasm for electric powered powered engineering grew with every undertaking.

It have turn out to be moreover inside the path of this time that Tesla commenced out to formulate his very very personal theories and inventions, frequently taking walks on them after his incredible art work hours. His relentless pursuit of information and experimentation become apparent in his off-hours, wherein he would hard paintings over calculations and sketches of capability improvements, which incorporates an early version of the induction motor.

This duration in Tesla's lifestyles have become pivotal not remarkable for the improvement of his technical data however additionally for the shaping of his character as an inventor. He displayed an unwavering determination to his paintings, a trait that would each useful resource and, at instances, restrict his destiny endeavors. The demanding situations he confronted, which includes the language barrier, financial constraints, and the strain to comply to current engineering practices, did now not deter him. Instead, they fueled his preference to innovate and push the limits of what modified into taken into consideration feasible in the place of electrical engineering.

Even despite the fact that they were only a short a part of Tesla's lengthy and fantastic career, his time in Europe had a sizeable impact. It modified into a period of improvement, reading, and the emergence of necessities that would sooner or later redesign the engineering and electric powered industries. One of the maximum notable brains of the nineteenth and 20th centuries emerged from those years, with studies that paved the manner for his subsequent accomplishments.

Tesla's region of innovation

Work at Continental Edison in Paris

Continental Edison in Paris

An important turning factor in Nikola Tesla's early profession changed into his revel in running at Continental Edison in Paris, which had a big impact on his next accomplishments. In 1882, Tesla left his function as an employee of the Central Telegraph Office in Budapest and traveled to

Paris, wherein he joined the Continental Edison Company.

During this era of speedy boom in the electrical vicinity, Tesla's get right of entry to at Edison's European subsidiary coincided with the fast unfold of electrical power and lighting fixtures systems at a few degree in the continent.

At Continental Edison, Tesla's function first of all concerned repairing and improving the business enterprise's electric tool. His paintings short examined his extremely good competencies as an engineer. Tesla's proficiency in managing electric powered system come to be now not quite a exquisite deal fixing system; it have turn out to be approximately expertise the underlying requirements and exploring techniques to decorate their performance and functionality. His early obligations covered putting in indoor incandescent lighting fixtures systems in Paris, a challenge that Tesla done with high-quality precision and creativity.

Tesla's time in Paris have become a length of profound expert boom. He gained sensible experience in the layout and control of dynamo machines and electric powered systems. This palms-on experience come to be valuable, grounding his theoretical knowledge and modern ideas in sensible fact.

The dynamo, which converts mechanical energy into electric powered powered strength, in particular interested Tesla. He observed functionality for improvement and commenced out to conceptualize adjustments that could enhance its overall performance, mind that would later come to be critical in his improvement of alternating modern-day-day (AC) systems.

Tesla's tenure at Continental Edison moreover exposed him to the wider elements of the electrical enterprise, from set up and safety to dealing with clients and knowledge the commercial organisation component of electrical businesses. This enjoy changed into important in shaping his statistics of the marketplace dynamics and realistic stressful situations of implementing electrical era. Tesla's art work ethic and progressive mind did now not circulate disregarded. His superiors fast identified his abilities, leading to opportunities to paintings on extra tough duties or even a rate to paintings on the improvement of the primary electric powered powered energy station in Strasbourg, France.

Edison continental (1882 - 1885)

This project changed into especially essential in Strasbourg. It modified into Tesla's first substantial threat to apply his talents on a huge-scale mission, and he tackled it with his signature mixture of intellectual records and modern practicality. In addition to the technical necessities of building the energy plant, Tesla's paintings in Strasbourg covered hassle fixing and enhancing the machine's efficiency and design.

More than only a artwork revel in, Tesla's time at Continental Edison served as a formative revel in that stepped forward his capabilities, broadened his records, and cemented his popularity as a colourful and resourceful engineer. Notwithstanding his achievements, Tesla have become extra conscious of the drawbacks of the then-dominant direct modern-day (DC) technology. His ideas about the opportunities of alternating modern started out to take shape, paving the manner for his next pastimes and the remaining advent of the AC device that might without a doubt rework the electrical agency.

With his departure from Continental Edison, Tesla's time in Europe got here to an cease and his voyage to America started. Carrying a letter of advent to Thomas Edison, Tesla sailed for New York in 1884, entire of ideas and a sturdy preference to growth his art work on alternating modern. Not handiest become this a shift in vicinity from his adventures in Europe to his travels in America, however it moreover marked a pinnacle turning issue in his profession, as he would fast be most important the charge within the maximum significant technological breakthroughs of his generation.

Chapter 3: First Innovations And Ideas

The formative years of Tesla's progressive brilliance had been spent in Europe; it have become currently that the roots of his eventual floor-breaking improvements had been planted. Following his graduation, Tesla commenced his expert career, first running on the Central Telephone Exchange in Budapest, Hungary. Here, in 1881, changed into in which his precise gift first began to reveal signs and symptoms of improvement.

Tesla's first professional profession

While in Budapest, Tesla conceived the concept of a rotating magnetic area, a precept that would later become crucial within the operation of alternating modern (AC) motors. This second of perception occurred all through a stroll in a metropolis park with a friend, as Tesla recited lines from Goethe's "Faust." The readability of his imaginative and prescient modified into such that he right now drew diagrams in the sand to illustrate the idea. This concept, progressive in its simplicity and splendor, marked a pivotal moment in Tesla's intellectual adventure and set the level for his destiny achievements.

Tesla joined the Continental Edison Company after relocating to Paris in 1882, wherein he acquired real-worldwide understanding inside the swiftly developing subject of electrical engineering. His paintings basically consisted on refining the format of direct contemporary (DC) turbines and debugging issues. Tesla obtained sensible enjoy and elevated his comprehension of electrical systems in the course of his tenure at Continental Edison, which have become surprisingly beneficial. It additionally bolstered his idea that alternating present day-day emerge as better to direct current-day, that may later motive him to disagree with Thomas Edison and precise electric powered agency figures.

Tesla's work on spinning magnetic scenario gadgets, which includes induction motors, become each different remarkable factor of his time in Paris. Not content material cloth to stop at electric powered engineering, he additionally experimented all through this time with radio waves, X-rays, and wireless energy transmission—subjects that would occupy a big portion of his destiny art work. For Tesla, a global powered thru smooth, green, and wi-fi power started out with these early assessments.

The rotating magnet

Even if his early studies confirmed promise, Tesla's European buddies regularly regarded his theories with suspicion. Because direct modern-day-day techniques were so extensively famous in the engineering and medical groups, Tesla's ideas have been considered radical and challenged conventional awareness. Tesla subsequently made his manner to the usa due to this competition as well as his developing preference to show off his discoveries on a bigger scale.

Thus, Tesla's years in Europe had been a time of exquisite intellectual growth, characterised with the aid of each novel discoveries and the problems of defying commonplace scientific records. During this era, he produced thoughts and inventions that solidified his characteristic as one of the maximum revolutionary inventors of his day and supplied the muse for his later successes. These were pivotal years in Tesla's career, as he would possibly later widely recognized. It have become at some point of this era that his innovations started to take form and his destiny as an inventor began out to take shape.

Moving to America

The Journey to New York

New York in 1884

Like the road he could have a take a look at within the years to come, Nikola Tesla's voyage to America, which signaled a pivotal 2nd in his existence and profession, changed into in addition dramatic and complete of uncertainty. With only his notable intelligence, a e-book of reference letters, and a imaginative and prescient of an alternating modern-powered destiny, Tesla sailed for New York City in 1884. For the younger inventor, who've grow to be nevertheless in large factor unknown out of doors of Europe but having already installed a popularity in some areas of the continent, this come to be a unstable bypass.

For Tesla, the transatlantic ride changed into more than first-class a physical journey—as an alternative, it became a plunge into the unknown and an journey right right into a universe of opportunities. The voyage become symbolic of the era's spirit of exploration and creativity. Even even though it come to be a common tale for masses immigrants on the time, Tesla's oceanic crossing held deep emotional that means. It represented no longer satisfactory the crossing of geographical limitations but additionally the passage from the antique to the modern-day global, from regular norms to novel principles.

When Tesla arrived in New York City, he placed a hectic, rapid-paced city that turned into very one in all a type from the calm, almost medieval cities he had grown up in. The metropolis end up a far cry from his preceding surroundings, with its pulsating strength, tall homes, and crowds of human beings from all walks of lifestyles. Tesla ought to come across his largest barriers further to his most fantastic successes on this dynamic placing.

Tesla had a few hard times during his first few days in New York. He had just a few pennies in his pocket whilst he arrived, which aptly captured the contrast in his lifestyles—a sparkly thoughts up towards the hard facts of existence. But he remained resolute. Tesla's first task have grow to be acclimating to his new domestic's complex social and expert environment. Armed with a letter of advent to Thomas Edison, one of the maximum famous inventors in America on the time, he looked for paintings. This letter stated, "I apprehend extraordinary men and you are without a doubt one in every of them; the alternative is that this more younger man." It become allegedly sent thru one of his former employers.

With Tesla's introduction to Edison, his profession entered a brief but momentous phase. Tesla grow to be first given a feature in Edison's organisation, which come to be already famous for its work on direct current (DC) electric systems. Though it can have seemed like a lucky damage, this hazard certainly paved the way for a complicated and hard partnership the numerous two terrific brains.

Tesla's opinions in New York at this early diploma of his American adventure organized him for his next innovations. The city become the appropriate putting for his grandiose goals due to its unexpectedly growing technical landscape. Here, Tesla began out to completely expand and sell alternating current-day (AC), a technology he felt would possibly virtually remodel the production, distribution, and usage of electricity.

Tesla with king peter II of Yugoslavia

It is not viable to magnify the significance of Tesla's relocation to America. He could allow his progressive juices drift in this united states of america of possibility, making contributions to era that might rework era forever. Not best have become his trip to New York a bodily pass, but it moreover marked the start of a destiny he passionately and truly observed. It prepared the target audience for his epic feats and the setbacks and victories that formed his awesome life.

Meeting Thomas Edison

Along together with his little possessions, Nikola Tesla delivered with him a imaginative and prescient that would in the future redecorate electric engineering whilst he immigrated to America in 1884. After receiving a letter of introduction from Thomas Edison's former employee and companion Charles Batchelor, Tesla turn out to be rapid moving into the busy Edison Machine Works offices in New York City. This assembly with Thomas Edison, a man whose name modified into already synonymous with American invention and organization, signaled the start of a short however momentous financial smash in his lifestyles.

Niola Tesla and Thomas Edison

Known for his improvements of the phonograph and the incandescent lightbulb, Edison end up a titan of the financial global. He have become an recommend of direct present day (DC) power, which turn out to be the norm within the US at the time. Contrarily, Tesla changed into an ardent recommend of alternating modern-day (AC), which he felt have emerge as the first-rate approach for transmitting power over extremely good distances. Their basically fantastic views on technology created the conditions for a complicated and often irritating partnership.

Tesla changed into first of all stimulated through the realistic technique of the American inventor, Edison, when they first met. As a self-taught inventor, Edison frequently forwent theoretical foundations in desire of trial-and-mistakes experimentation and as an alternative relied specially on empirical methods. Tesla, but, changed proper into a man of idea, firmly based absolutely in mathematical calculations and medical mind. Notwithstanding those disparities, Edison saw Tesla's capability and taken him on board to paintings on enhancing the DC producing centers.

Tesla had a difficult and illuminating experience at the identical time as jogging at

Edison Machine Works. He hooked up masses of beyond everyday time, frequently walking from dawn till a ways after middle of the night. His maximum essential responsibility turn out to be to discover answers for the issues with Edison's DC mills and distribution tool. Tesla made some of upgrades recommendations, but Edison frequently overlooked them as they dealt primarily with AC systems, which Tesla categorically denied.

A famend story from Tesla's days jogging with Edison problems a promised incentive. Edison supplied Tesla $50,000 if he also can need to correctly remodel the inefficient DC mills, likely in comic story or as a venture. After months of hard paintings, Tesla came up with a technique to this excellent task. But Edison overlooked his request for the promised cash, joking, "Tesla, you do not apprehend our American humor." Shortly after, feeling deceived and disheartened, Tesla introduced his resignation.

Although Tesla and Edison had not been proper now related, this event only signaled the begin of a prolonged and illustrious competition. After advocating for AC technology, Tesla ultimately teamed up with George Westinghouse, the primary rival of Edison. The warfare that resulted between AC and DC, dubbed the "War of Currents," went at once to turn out to be a pivotal 2nd within the development of electrical engineering.

Looking returned, Tesla's short partnership with Edison was a turning 2nd in his expert life. He advanced his abilties, subtle his theories, and solidified his faith within the possibilities of alternating cutting-edge in some unspecified time in the future of this time. Tesla's visionary aspirations and Edison's pragmatic processes contrasted sharply, highlighting no longer nice the variations within the pathways taken thru the two excellent innovators but moreover laying the inspiration for Tesla's future accomplishments that could ultimately convey in a brand new technology of electrical era.

Chapter 4: George Westinghouse

Tesla and Edison

When Nikola Tesla arrived in the United States in 1884, the narrative of his encounter with Thomas Edison—one of the most charming moments of his existence—began. With little extra than a letter of recommendation from Charles Batchelor, an partner and previous business enterprise of Edison, Tesla arrived in New York City full of expectations. The letter became written to Edison and guarded a strong advice: "This more youthful guy is one of the exquisite men I understand; you are the opportunity."

Tesla and Edison's initial assembly became a turning issue within the records of electrical improvement as well as for Tesla. Direct present day (DC) electrical use became invented thru Edison, who changed into already well-known as an inventor and businessman. On the alternative hand, Tesla supported AC, or alternating modern, considering that he perception it turn out to be better for transmitting power over exceptional distances. Edison employed Tesla on the Edison Machine Works after identifying his capacity, however their divergent reviews.

Nevertheless, the partnership modified into difficult and short-lived. Tesla end up entrusted with reworking Edison's inefficient motors and turbines in addition to working on improving the DC turbines. An offer to compensate Tesla for huge upgrades in his DC dynamo designs is said to were made thru Edison, who's well-known for having promised to pay $50,000. Tesla placed up a great deal of strive and considerably advanced the Edison designs. But at the same time as Edison confirmed his paintings, he disregarded the pledge as an American shaggy canine story, and their partnership ended bitterly.

Disillusioned and feeling underappreciated, Tesla give up Edison's enterprise. Not first-class did this parting sign the give up of their quick partnership, but it moreover laid the idea for the imminent "War of Currents." Tesla and Edison, with their respective AC and DC electricity structures, can also want to later face off in a opposition to control the development of electrical power structures inside the destiny.

Though brief and turbulent, Tesla's relationship with Edison have become crucial in figuring out his destiny. During this time, Tesla's strength of thoughts to reinforce alternating contemporary have emerge as more potent, placing him at the path that might in the end result in some of his best accomplishments. Along with being exposed to the cruel realities of the enterprise organization worldwide, Tesla's enjoy additionally made him recognize how idealistic and purist his personal technique to technological understanding and advent emerge as.

Despite disagreements and broken ensures, the Tesla-Edison partnership is despite the fact that regarded as a pivotal period within the records of era. It draws hobby to the collision of first rate minds, each with a totally specific outlook at the course that electricity will take. It grow to be a crucial event for Tesla, one that bolstered his self-control and prompted his future pursuits within the field of electrical engineering.

The War of Currents

Alternating Current vs Direct Current

The War of Currents turn out to be a critical length in the records of electrical power that concerned a bitter conflict among opposing electric powered structures: Direct Current

(DC), supported with the beneficial resource of Thomas Edison, and Alternating Current (AC), supported through Nikola Tesla. The battle modified into now not simplest a technical argument; it additionally worried a warfare of ideologies, commercial organization processes, and private grudges that had a big effect on how electric powered electricity have come to be disbursed within the destiny.

The primary difference among AC and DC have become on the center of this argument. Early electric powered powered distribution structures used direct modern-day, which flows constantly in a unmarried route and end up promoted via Edison. Due to its ease of use and ease, it have end up as quickly because the primary technique of presenting strength to towns. The astute businessman and inventor Edison had made big investments in DC infrastructure and era, constructing energy plant life to mild up places inclusive of New York. DC, but, had some of drawbacks, most significantly the dearth of ability to move statistics successfully over long distances without suffering excessive energy loss. This intrinsic drawback made it impractical and costly to distribute strength substantially as it necessitated the development of strength plant life at near distances from each exceptional.

Contrarily, Alternating Current, of which Tesla emerged as the most distinguished and creative propose, alternates direction on a everyday basis and is with no problem convertible to certainly one of a kind voltages via using transformers. Because of this option, AC have become quite powerful in transmitting data over extremely good distances, minimizing energy loss and allowing the use of fewer electricity vegetation to serve wider areas. With his high-quality understanding of electromagnetic fields, Tesla become capable of every envision and create the strategies for completely expertise the potential of AC. His cutting-edge introduction of the induction motor and transformers, alongside along along with his guide of the polyphase gadget, proved that AC have end up the more effective strength distribution method for huge-scale and lengthy-distance applications.

The War of Currents peaked whilst Tesla and fellow entrepreneur and inventor George Westinghouse engaged Edison in a sour and frequently public conflict. Edison commenced out a marketing campaign to depict AC as volatile and unreliable so that you can undermine the generation. As a part of this advertising advertising campaign, animals have been electrocuted in public to consciousness at the claimed dangers of AC. Driven by means of way of manner of the notion that AC changed into now not first-rate regular but moreover the manner of the future for electric powered power, Tesla and Westinghouse persisted in spite of these strategies.

Tesla's AC Current

The building of Niagara Falls' hydroelectric power plant was the turning aspect in this war. The settlement to harness the massive energy of the falls changed into provided to Tesla and Westinghouse, who used AC to hold electricity at some stage in a distance of greater than twenty miles to Buffalo, New York.This accomplishment marked a sea exchange through demonstrating the great protection and effectiveness of AC. After that, AC quickly rose to prominence as the enterprise famous for dishing out electricity, a feature it nonetheless holds these days.

Beyond truly being a technological competition, the War of Currents marked a turning component in American industrialization via manner of influencing the manufacturing, distribution, and use of strength. During this time, Tesla's brilliance and imaginative and prescient had been emphasized, however it also tested how era, commercial interests, and man or woman personalities have interaction to form scientific development. Ultimately, AC's triumph served as proof of Tesla's electrified international imaginative and prescient, which has had a profound effect on contemporary lifestyles.

Collaboration with George Westinghouse

The partnership among Nikola Tesla and George Westinghouse marked a turning detail within the records of the War of Currents, an generation-defining battle that would decide the path of electrical electricity inside the destiny. The alternating present day (AC) device emerge as capable of defeat the direct contemporary (DC) techniques that Thomas Edison and his supporters were pushing because of the fact to this collaboration, that have emerge as characterized with the resource of mutual recognize and a commonplace purpose.

Early on, Tesla's AC device's promise changed into diagnosed via producer and inventor George Westinghouse. Being unrestricted through previous investments in DC power structures, Westinghouse turn out to be greater open to Tesla's ground-breaking requirements than Edison was.

Thomas Edison DC Current

Tesla's innovations have been the cornerstone of the Westinghouse Electric Company's approach to impress America, which noticed the agency created in 1886 grow to be a effective force within the electric powered region.

Westinghouse found out that Tesla's AC technology have become more steady and extra fee-powerful to put in on a large scale, in addition to being extra green over prolonged distances. But the road to demonstrating AC's dominance wasn't without its problems. As a powerful opponent and public members of the circle of relatives professional, Edison released a decided advertising advertising marketing campaign to denigrate AC via characterizing it as unstable and untrustworthy. Throughout this advertising and marketing and marketing marketing campaign, animals had been publicly electrocuted with AC energy for you to instill worry and skepticism within the public.

Westinghouse and Tesla stayed robust inside the face of these difficulties. Ever the showman and innovator, Tesla performed open demonstrations to interest on the efficacy and safety of AC. A pivotal second happened finally of an instance of this type at some stage in the 1893 Chicago World's Fair, wherein Tesla and Westinghouse used AC power to mild the complete carnival. This showcase no longer simplest amazed onlookers and commercial company specialists, however it furthermore proved AC's supremacy in a real-global, convincing manner.

The constructing of the number one big-scale AC strength plant at Niagara Falls end up the top of Tesla and Westinghouse's partnership. When it turn out to be completed in 1895, this engineering surprise served as a tribute to Westinghouse's spirit of entrepreneurship and Tesla's vision. The War of Currents come to be efficaciously located to an give up via way of the Niagara Falls energy plant's success, which made AC the everyday popular for the transmission of electrical electricity.

Westinghouse and Tesla had many monetary setbacks during their collaboration. Westinghouse as soon as had to renegotiate royalties with Tesla as it come to be regularly coins-strapped because of the full-size charges of competing with Edison and building the infrastructure for AC electricity. Tesla is rumored to have torn up his royalty agreement in an illustrious and perhaps fictitious act of sacrifice for the more suitable. While this circulate preserved Westinghouse's organisation, it located Tesla in a risky monetary function for the the relaxation of his lifestyles.

Niagara Falls

More than only a employer alliance, Tesla and Westinghouse's partnership represented the union of notable minds that complemented each specific. Tesla contributed ground-breaking thoughts and brilliance in generation, while Westinghouse contributed abilties in sensible company control and a flair for navigating the industrial terrain. Collectively, they now not simplest shifted the route of electrical generation however additionally set up the framework for the current energy device, having an effect on billions of humans international.

The dating amongst Tesla and Westinghouse has left a protracted-lasting legacy that serves as a compelling instance of the manner creativity, hard art work, and endurance can overcome vested interests and exchange the arena. Not most effective changed into AC a hit over DC in the War of Currents, but it additionally proven the strength of ingenuity.

The Chicago World's Fair and the Niagara Falls Project

The 1893 World's Columbian Exposition, moreover called the Chicago World's Fair, have become a exquisite theater in the War of Currents, which emerge as characterised by means of manner of the acute opposition amongst direct current-day (DC) and alternating contemporary-day (AC) structures. This event modified into a watershed within the recognition and application of alternating modern-day for the distribution of electrical electricity. It additionally showed the genius of Nikola Tesla and his companion, George Westinghouse.

The cause of lighting up the whole fairground served because the centerpiece of the exposition and represented the start of a new age in era. Thomas Edison and General Electric, who were ardent supporters of direct modern-day, fought difficult for the agreement, but in the end, Westinghouse acquired often because of their lower provide and Tesla's floor-breaking AC technology. This triumph represented a trade inside the path of technical development and become extra than only a commercial fulfillment.

The 1893 World's Colombian Exposition

It have become no longer whatever quick of a technological wonder whilst Tesla's alternating present day-day device have turn out to be located into use on the Chicago World's Fair. The carnival changed into illuminated through over two hundred,000 lightbulbs, impressing guests with an tremendous and unmatched feat of electrical electricity. In addition to being greater inner your way and inexperienced than Edison's DC approach, Tesla's AC device grew to turn out to be out to be greater secure and extra useful for lengthy-distance electricity transmission. By proving the prevalence of alternating contemporary-day and starting up the door for its wider use, the success on the exposition demonstrated Tesla's efforts.

After the success on the World's Fair in Chicago, Tesla and Westinghouse set out on a miles extra bold task: using Niagara Falls' power to create power. A demonstration of Tesla's idea to use the power of nature to create sustainable strength have come to be the Niagara Falls Power Project, which modified into meant to be the primary large-scale hydroelectric electricity station in facts.

The assignment comprised constructing a huge electricity plant that used Tesla's alternating cutting-edge-day approach to convert the water's kinetic electricity into electric electricity. This undertaking wasn't with out its problems. It pushed the boundaries of contemporary generation and required a notable deal of ingenuity in electric powered powered engineering and turbine format.

An crucial turning factor within the development of power changed into reached with the a achievement finishing touch of the Niagara Falls Power Project in 1896. The functionality to generate and switch considerable portions of electrical electricity all through wonderful distances modified the way energy changed into produced and used for the primary time. The mission's fulfillment cemented alternating modern's hegemony and located Tesla and Westinghouse as leaders in the electric powered powered engineering region.

More than just engineering feats, the Chicago World's Fair and the Niagara Falls Project served as showcases for Tesla's brilliance and vision. These accomplishments no longer incredible helped alternating modern win the War of Currents, however additionally they set up the framework for the current electric powered grid. Tesla made progressive achievements in some unspecified time in the future of this time that changed the course of human records all of the time and ushered in a trendy technology of technological increase. His contributions had been not in reality technical.

Chapter 5: Inventions And Discoveries
The Tesla Coil and Wireless Transmission

Nikola Tesla's maximum well-known innovation, the Tesla Coil, is a tribute to his understanding inside the realm of electromagnetic and represents a crucial turning aspect in the facts of electrical engineering. Not wonderful modified into this great gadget an invention, however it also represented a sizeable development in the statistics and control of electrical currents, some issue that Tesla started out out going for walks on in the overdue 1800s. The Tesla Coil is a resonant transformer circuit used to generate immoderate-frequency, low-voltage, excessive-voltage alternating modern. It is a not unusual subject matter in medical training and a illustration of Tesla's inventiveness and imaginative and prescient because of its capacity to supply wonderful electric powered arcs and its utility in wi-fi transmission assessments.

Tesla's steadfast religion inside the capability of alternating contemporary (AC) systems delivered about the advent of the Tesla Coil. His preceding trends in AC power systems, which had already prepared the foundation for the War of Currents with Edison's direct cutting-edge (DC) systems, led him to begin artwork on the coil. The Tesla Coil, however, represented a bold foray into the undiscovered realm of immoderate-frequency electric powered phenomena in preference to definitely an development over previous technology.

Tesla's hassle with wi-fi strength transmission served because the number one motivation within the again of his artwork on the Tesla

Coil. In his imaginative and prescient, there may be no wires or connections ; alternatively, a international linked via an unseen network of energy. These checks relied carefully on the Tesla Coil, which produced relatively high voltages able to theoretically transmitting energy wirelessly over wonderful distances. To make this vision a reality, Tesla experimented with the coil in his laboratory in New York after which at his Wardenclyffe Tower assignment.

Bipolar Tesla transformer in 1908

The primary and secondary coils of a Tesla coil are each associated with a separate capacitor in order for the coil to characteristic. In order to resonate at the equal natural frequency, the two coils and capacitors have been tuned. A magnetic difficulty is produced at the same time as an alternating present day is executed to the primary coil, and this induces a modern-day to glide thru the secondary coil. The Tesla Coil can produce specially excessive voltages due to the truth to a mechanism known as resonant inductive coupling.

Amazing matters occurred in the course of Tesla's Tesla Coil demonstrations. He have to regularly carry out in public, lighting fixtures wi-fi lamps and wowing crowds with lightning-like arcs. These displays weren't just suggests; they served as proof of concept for his theories on wi-fi transmission and his extra entire plan for the improvement of strength.

The sensible use of the Tesla Coil for wireless power switch faces many obstacles no matter its capability. The most essential of these have become the inefficiency of prolonged-distance strength switch, which Tesla struggled with however changed into by no means able to virtually treatment.However, the Tesla Coil's demonstration of key standards set the level for similarly improvements in wireless verbal exchange, radio technology, or maybe clinical tool.

Today, the Tesla Coil is respected as a instance of the limitless potential of clinical inquiry in addition to a mark of Tesla's inventiveness. Even regardless of the reality that the Tesla Coil's usefulness for wi-fi transmission turn out to be restricted, it despite the fact that represents a huge milestone in the statistics of electrical engineering and suggests Nikola Tesla's unwavering faith in the capacity of alternating present day similarly to his unrelenting quest to create an interconnected universe powered through invisible power threads. The arcs and sparks of the Tesla Coil, which constitute Tesla's legacy, never surrender to amaze and excite us. They function a consistent reminder of the price of imagination and the unwavering pursuit of innovation.

Radio and Remote Control

Within the outstanding catalog of Nikola Tesla's enhancements, the radio and far off manage are considered foundational portions that show his vision and unmatched ability to assume past the policies of his technology. In addition to developing the muse for cutting-edge wireless communication, Tesla's contributions in numerous areas converted how humans have interaction with devices and technology.

Tesla's remote control board

Following Heinrich Hertz's groundbreaking research, Tesla's fascination in electromagnetic waves led him to pursue a profession in radio. The concept of the use of the ones waves for wireless communique captivated him. Tesla initially described his concept for a tool for the wi-fi transfer of energy and records in a speech given in London in 1892 to the Institution of Electrical Engineers.He optimistically stated a destiny wherein data, music, or maybe pictures might be sent wirelessly at some point of awesome distances in his later lectures within the United States, increasing on this vision. In addition to being modern, Tesla's theories were amazingly in advance of his time, foreseeing the technology of world communique properly earlier than it materialized.

Tesla performed for multiple patents in 1897 that covered the factors of a wireless verbal exchange tool, collectively with an electromagnetic wave transmission and reception mechanism, an induction coil with the ability to supply high-frequency alternating currents, and a manner to regulate the frequency of these currents. These patents are these days called being critical to the improvement of radio. Tesla, but, saw radio not simply as a tool for communication but additionally as a platform for power transmission through wireless generation. He anticipated a society wherein electricity became without a doubt as interconnected as data.

Tesla's artwork on faraway control changed into probably plenty extra modern. He stunned Madison Square Garden crowds in

1898 by way of showing off a tiny, radio-controlled watercraft during an electrical show. With the assist of a tiny antenna, this boat come to be able to circulate left, proper, or in advance or perhaps flash its lighting fixtures in response to guidelines from Tesla transmitted with the beneficial aid of a transmitter. This presentation served as greater than handiest a show; it turn out to be a display off for the usefulness of the use of radio waves to remotely manipulate machine. Even despite the fact that the era Tesla displayed turn out to be primitive via current requirements, it changed into not something brief of wonderful at the time, placing the premise for the entirety from state-of-the-art army drones to far flung-managed toys.

On the alternative hand, reactions to Tesla's contributions to radio and a ways off manage were a aggregate of amazement and distrust. Some praised him as a genius, at the identical time as others doubted the mind' applicability and protection. Furthermore, he clashed with other innovators due to his paintings in the ones areas, maximum extensively Guglielmo Marconi, who may also want to move without delay to be credited with developing radio. After an extended criminal combat over the improvement of radio, Tesla changed into ultimately honored for his efforts posthumously, with the U.S. Supreme Court preserving his patent rights in 1943.

Tesla's efforts with radio and a long way off control display each his brilliance and his unwavering vision of a international in which era could likely dismantle limitations and open up new opportunities. These technological advances have been more than just products of their era; they provided us a startlingly smooth glimpse into the future that Tesla believed in, one wherein energy and statistics could be transported thru the air, uniting and empowering people in strategies that had in no way earlier than been notion viable.

Contributions to Electrical Engineering

The present electric powered international has been notably inspired by means of Nikola Tesla's profound and an extended way-sporting out contributions to electric engineering. Not most effective became he a first-rate truth seeker even as it came to modern mind, however he end up furthermore an unrivaled artist at bringing those thoughts to existence. The important thoughts in the lower returned of a big part of our current electric infrastructure and generation have been installed with the aid of Tesla's efforts.

Tesla's invention of alternating contemporary (AC) systems is his maximum great contribution to electric engineering. Although direct present day (DC) become the norm inside the overdue 1800s, Tesla turn out to be aware of DC's drawbacks while it came to prolonged-distance electricity transmission. His invention of transformers and AC induction motors made it feasible to generate strength at power flora and supply it across extremely good distances with little loss.This innovation substantially altered the distribution and usage of electricity, improving its overall performance and accessibility for full-size use.

One of his different remarkable inventions, the Tesla coil, illustrated the opportunity of wi-fi energy switch. This immoderate-frequency, high-voltage transformer served as a prototype for contemporary radio generation and will produce notably effective electric powered fields. Even even though it became no longer clearly carried out in his lifetime, Tesla's concept of wi-fi electric power transfer set the inspiration for later wireless communique generation.

Furthermore, Tesla made a significant contribution to the arena along along along with his artwork on polyphase structures. Power generation, transmission, and motor operation have become more green manner to his innovation of the 2-phase electric powered energy system.A vital thing of present day electric engineering, this machine uses numerous alternating currents which might be out of section with each special to facilitate the green operation of electrical machinery and power grids.

Tesla found novel electric powered phenomena due to his excessive-frequency cutting-edge-day investigations. He investigated the opportunity of excessive-frequency currents for therapeutic features, which resulted in upgrades in electrotherapy. Even if some of his theories on this discipline had been theoretical, they recommended extra studies into the uses of power in treatment.

Tesla moreover confirmed brilliant foresight in predicting the adoption of renewable energy belongings. Long earlier than there were the technological gear to perform that, he hypothesized about harnessing the strength of the sun, wind, and excellent renewable assets.In this revel in, his imaginative and prescient foreshadowed the modern look for renewable power property.

Wireless Electricity

Many advances in electric engineering may be attributed to Tesla's theoretical research and patents. His passion for developing and refining electric systems determined out an unshakable belief in the capability of generation to decorate humankind. Tesla's legacy in electric engineering is marked through his good sized achievements which have persevered the check of time, notably impacting the way we produce, transmit, and devour electric powered powered electricity, in spite of confronting many limitations and disappointments. In addition to advancing the electric industry, his work served as an perception for future generations of engineers and inventors to hold pushing the frontiers of electrical research.

The Colorado Springs Period

Experiments with High-Frequency Currents

A first-rate turning aspect in Nikola Tesla's development as an inventor and scientist have emerge as the Colorado Springs Period. Following his turbulent time in New York, Tesla searched for a present day place to live that might provide him the seclusion and resources required for his ambitious experiments with excessive-frequency currents and wi-fi transmission. Colorado Springs provided the proper putting for Tesla's bold initiatives due to its massive open areas and relatively untouched topography.

When Tesla moved to Colorado Springs in 1899, he installation a facility wherein he have to perform a number of his maximum crucial studies. This lab's predominant characteristic have turn out to be a large Tesla coil, that could produce distinctly excessive frequencies and voltages. The scale and skills of this gadget have been in evaluation to a few issue that had ever been constructed. Tesla's idea of wireless communication and power transmission served as the impetus for his trials in Colorado Springs. He idea that prolonged-distance wireless transmission of energy or even messages might be feasible with the ideal era.

During this time, Tesla's work with immoderate-frequency currents become one of the most essential research fields. He driven the limits of what turned into then taken into consideration scientifically viable by using investigating the conduct of these currents at hitherto unexplored depths. Tesla's tests showed that robust electric powered discharges can be produced with high-frequency currents; these discharges is probably seen arcing at some stage in his lab. Not only had been those indicates putting to check, however in addition they marked a jump ahead in our statistics of the opportunities of electromagnetic radiation.

Tesla completed experiments that went beyond precept. He often placed himself in danger to thoroughly verify his theories. At one point, he showed the opportunities of wireless electricity transmission through famously wirelessly lights vacuum tubes on the same time as keeping them in his palms and looking them glow brightly. Since Tesla turn out to be unhurt, this experiment have emerge as not high-quality a proof of concept but moreover a lovely demonstration of the feasible protection of excessive-frequency currents.

Moreover, on the identical time as he changed into at Colorado Springs, Tesla discovered uncommon electric powered phenomena. He notion that the alerts he modified into deciding on up have been coming from area and speculated that they is probably attempts at touch from other worlds. Even despite the fact that these assertions have been taken into consideration with skepticism on the time, they served as some different proof of Tesla's innovative wondering and unwavering quest for understanding.

In Colorado Springs, Tesla moreover labored on tasks concerning tries to alternate energy thru the Earth, a principle he dubbed "Earth resonance." He postulated that electrical energy is probably transmitted to any place on Earth with the useful aid of harnessing the planet's inherent frequencies. His studies in this concern impacted mind in earthquake detection and evaluation in addition to later theories and applications in geophysical studies.

Chapter 6: Colorado Springs Laboratory

While Tesla saw giant creativity and discovery within the course of the Colorado Springs Period, there have been drawbacks as nicely. Both technical and financial limits regularly hindered his paintings. His meager price range had been pushed thru the large strength necessities of his studies, which occasionally brought approximately disagreements with the close by electricity agency.

Notwithstanding the ones troubles, Tesla's tenure in Colorado Springs serves for example of his inventiveness and unwavering willpower to the boom of era. During this time, he achieved experiments that superior our knowledge of wi-fi transmission, immoderate-frequency currents, and electromagnetic fields. They played a pivotal characteristic in the improvement of numerous cutting-edge-day technical innovations, which include as wireless power transfer and radio communiqué. At addition to showcasing his extremely good creative potential, Tesla's art work at Colorado Springs solidified his popularity as a trailblazer inside the undertaking of electrical engineering and a visionary properly in advance of his time.

Theories and Experimentation

Between 1899 and 1900, Nikola Tesla lived in Colorado Springs. During this time, he completed amazing experiments and advanced floor-breaking theories, in particular inside the project of Wi-Fi verbal exchange. Tesla emerge as not satisfactory interested by wi-fi technology as a systematic company; he furthermore determined it as a way of organizing hitherto unthinkable global connections. He positioned the seclusion and unspoiled surroundings of Colorado Springs, which allowed him to perform his immoderate-voltage, immoderate-frequency experiments on a scale he had in no manner tried earlier than.

Tesla looking for of Colorado springs lab

Based on his earlier assessments in New York, Tesla persevered and considerably amplified them in Colorado Springs. Here, he built a facility that might preserve a huge Tesla coil that would produce tens of thousands and thousands of volts of strength. This gadget changed into critical to his experiments in wi-fi transmission. Tesla's primary reason became to show that electrical electricity is probably transmitted wirelessly over prolonged distances, no longer truly as a systematic interest, but as a sensible method of communiqué and electricity transfer.

One of the maximum superb experiments performed in the course of this era come to be the illumination of wireless lamps. Tesla set up that he ought to mild lamps full of a fuel, together with phosphor-blanketed tubes or bulbs, without any wires, at distances of up to 25 miles from their strength deliver. This success changed into now not high-quality a testament to his information of electromagnetic fields but moreover a glimpse proper right into a capacity future wherein power can be transmitted through the air, powering gadgets remotely.

Tesla moreover achieved experiments that contributed to the improvement of radio technology. He despatched electromagnetic waves thru the floor and the environment, detecting them at splendid distances. These experiments laid the concept for the later improvement of radio communique, regardless of the reality that Tesla's hobby remained at the transmission of energy in region of indicators.

Another groundbreaking issue of Tesla's art work in Colorado Springs worried the Earth's ionosphere. Tesla hypothesized that the

Earth's environment might also want to conduct strength and may be used to transmit strength and information throughout huge distances. He theorized approximately the life of an electrical conductor immoderate above the Earth, which we now recognize because the ionosphere. His experiments in transmitting electrical electricity worried sending effective electrical pulses into the ground and measuring the electrical reaction at faraway elements, efficaciously the usage of the Earth as a conductor.

Tesla's Colorado Springs experiments were characterized via their ambitious scale. He recorded measurements of electrical phenomena that were outstanding, together with measuring the resonant frequencies of the Earth. Tesla believed that, via using the ones resonant frequencies, it'd be possible to transmit strength and messages throughout the globe with out wires.

But there have been some difficulties for Tesla inside the course of his live in Colorado

Springs. His experiments were so big-scale that they regularly interfered with the developing telegraph tool, which resulted in courtroom cases from nearby businesses and citizens. Furthermore, funding for Tesla's experiments changed right right into a continuous issue due to the reality they have been steeply-priced. This fairly effective segment came to an forestall with the 1900 closure of the Colorado Springs laboratory, no matter the fact that Tesla carried on alongside collectively with his imaginative and prescient of wireless conversation and energy transmission in subsequent ventures.

The Colorado Springs section gives evidence of Tesla's brilliance and his steadfast dedication to his undertaking. His theories and experiments at some point of this era revolutionized the region of electrical engineering and set the foundation for numerous contemporary generation. Although not certainly placed out all through his lifetime, Tesla's contributions to wireless communique revolutionized the manner electricity and statistics are transmitted, extensively converting the trajectory of technological development inside the 20th century and past.

The Colorado Springs Laboratory

Tesla determined on Colorado Springs in 1899 because of the truth he desired a more non-public and roomy placing for his excessive-voltage, immoderate-frequency experiments. This vicinity furnished him with the solitude and room he desired, similarly to the opportunity to make use of the neighborhood electric phenomena. Tesla desired to push the boundaries of what modified into then taken into consideration technological comprehend-how to similarly his studies in electric powered generating and wi-fi conversation.

As speedy as he were given to Colorado Springs, Tesla commenced building his laboratory. The facility changed into precise because it had a massive coil that could produce especially immoderate voltages, a characteristic that could come to be associated with his wireless energy research. This coil, moreover referred to as the "Magnifying Transmitter," changed into created to send electric strength in the course of long distances with out the need for wires. Tesla had a completely unique idea: to installation a international network for energy switch and wireless conversation.

Colorado Springs lab-resonance

Some of Tesla's maximum placing and bold experiments came about inside the Colorado Springs Laboratory. He surprised observers and supplied a glimpse into the future of wireless energy while he managed to ignite wireless lamps miles far flung from his energy supply. In addition, he created faux lightning bolts through his laboratory experiments, and there have been payments of incredible electric powered powered shows and thunderous noises that might be heard and visible from awesome distances. In addition to showcasing Tesla's scientific acumen, those experiments determined his profound statistics of the natural world.

But Tesla confronted many problems all through his stay in Colorado Springs further to successes. Large quantities of electricity have been regularly wished for his experiments, which brought on commonplace energy outages that aggravated the locals and corporations. Furthermore, Tesla's already fragile price variety had been strained with the useful resource of the expenses of coping with the form of huge commercial enterprise corporation. Motivated thru his conviction within the importance of his art work, Tesla persisted in the face of those obstacles.

Tesla proposed the idea that the Earth might be a conductor of resonant frequencies whilst he emerge as in Colorado Springs. He counseled that electrical electricity might be despatched to each location on Earth by the use of utilizing the planet's inherent resonance. Even if it wasn't definitely finished in Tesla's day, this concept set the degree for later upgrades in energy transmission and wireless communique.

Despite its short life, the Colorado Springs Laboratory served as a center of invention and a turning 2d in Tesla's professional life. He done some of his maximum crucial experiments here, which had a long-lasting impact at the domains of electrical engineering and wi-fi communications. The laboratory will constantly be related to Tesla's unwavering spirit of exploration and discovery, similarly to the scientific advances it witnessed. Tesla's tenure in Colorado Springs have turn out to be marked via the use of a extraordinary amount of invention and discovery notwithstanding overcoming technological and economic demanding situations, solidifying his reputation as taken into consideration one in all facts's most incredible and visionary inventors.

The Wardenclyffe Project

Vision of Global Wireless Communication

Wardenclyffe Tower

One of the maximum audacious and visionary obligations within the records of technological knowledge and technology is Nikola Tesla's Wardenclyffe Project. Tesla's imaginative and prescient of global wireless communique, which became an extended manner earlier of its time and nonetheless fascinates and evokes people these days, became on the center of this business agency.

Tesla's profound records of electromagnetic and his earlier achievements with wi-fi transmission served as the foundation for his vision. In his imaginative and prescient, there can be an invisible electricity community connecting all elements of the planet, allowing unrestricted communication throughout exceptional distances. The prototype station, Wardenclyffe, changed into alleged to be the primary on this international device, proving that wi-fi electricity and statistics transfer became feasible.

A superb tower at the middle of the Wardenclyffe fame quo come to be supposed to carry communications and, in a more audacious skip, to supply wireless strength over the Atlantic. Tesla postulated that he may additionally ship electromagnetic waves at some stage in notable distances with little strength loss with the resource of harnessing the Earth's ionosphere. His wireless worldwide changed into going to revolve spherical this tower due to its uncommon length and format.

Tesla had more bold intentions for Wardenclyffe than surely technological improvement. He taken into consideration this attempt as a manner to bridge geographic divides and unite individuals from everywhere inside the global. He envisioned a time in which information might be abruptly shared for the duration of oceans, together with information, communications, or maybe pix, connecting human beings and shrinking the globe. There were countless possible makes use of for it, such as broadcasting, navigation, remote getting to know, and real-time data updates.

But Tesla had a imaginative and prescient that went past technology to include profound humanitarianism. He concept that unrestricted access to energy and facts could promote global peace and cooperation on the identical time as helping inside the reduction of inequality. For him, Wardenclyffe represented improvement not first-class for the us however the entire globe.

Even although the idea grow to be first rate, the Wardenclyffe Project ran into masses of issues. Money problems had been an ongoing impediment. J.P. Morgan, Tesla's foremost backer, first gave the venture his blessing, but as Tesla's dreams prolonged, Morgan misplaced patience and hobby. Furthermore, opinions in the scientific network regarding the viability of Tesla's formidable desires differed. Some have been inspired thru his idea, however others were dubious about the era and viability of wireless energy transfer in this grand scale.

When World War I broke out, extra money and reputation have been directed some region else, which made matters even greater complicated. The task, which changed into already having financial issues, end up having increasingly more problem getting the money it preferred. The dream that Tesla had as quickly as regarded possible started to disintegrate.

Wardenclyffe Demolition

Ultimately, the Wardenclyffe Tower come to be demolished and the location left deserted, serving as a moving instance of unrealized capacity. Nonetheless, Tesla's goal for wireless communique round the arena lives on. The associated networks and immediate communique of the modern-day international are in masses of respects a manifestation of Tesla's imaginative and prescient. Although the era of in recent times is not just like that of Tesla's actual plans, the concept of a related global continues to be real to his vision.

Looking once more, the Wardenclyffe Project is a virtually great, if a hint too early, step into the destiny. The worldwide wireless network that Tesla anticipated changed into a forerunner of the contemporary technology that we take as a right. It is evidence of his brilliance and his steadfast religion inside the potential of technological know-how and technology to exchange the arena.

Chapter 7: Financial Struggles And The Morgan Relationship

The complicated connection among Tesla and financier J.P. Morgan, which began off promising but in the end resulted in unhappiness and disillusionment for Tesla, became on the center of these issues.

From his in advance experiments in Colorado Springs, in which he had done wonderful fulfillment in wireless transmission, Tesla derived the concept for Wardenclyffe. His audacious concept for the Long Island city of Wardenclyffe modified into to convey collectively a massive wi-fi communique network capable of sending signs to London and Paris throughout the Atlantic. Tesla observed Wardenclyffe as greater than most effective a method of conversation; it emerge as step one closer to his purpose of transmitting electric energy wirelessly.

J.P. Morgan

It changed into no smooth challenge to normal budget for a task this groundbreaking. Tesla sought the assistance of J.P. Morgan, a great financier of the generation, renowned for his ventures into developing era and sectors. At first, Morgan became enthralled with Tesla's concept and consented to contribute a huge amount to provoke the challenge. This funding established every Tesla's capability to persuade others and the viability of his floor-breaking concepts.

But because the undertaking advanced, Tesla's financial issues grew more immoderate. Much more money than anticipated became needed to collect the

Wardenclyffe tower and centers. Due to Tesla's excessive interest to detail and his want for perfection, there had been severa redesigns and adjustments, which drove up prices even in addition. He located that he end up constantly combating to get extra cash, not without a doubt to finish the project however furthermore to hold his studies and dwelling costs.

Morgan's first of all resolute help started out to falter due to the fact the mission ran into delays and developing costs. Tesla had now not added on his guarantees of a modern verbal exchange gadget, and the a hit transatlantic wireless transmission by means of the usage of Marconi in 1901—the usage of a miles much less complicated and less steeply-priced method—will increase questions about Wardenclyffe's usefulness and profitability. Morgan commenced out to doubt the mission's monetary sustainability since he have become greater inquisitive about the viable business returns.

Ever the optimist and visionary, Tesla made an try to influence Morgan and precise feasible buyers of the mission's viability, frequently exaggerating Wardenclyffe's capabilities and its instant opportunities. But increasingly humans have been beginning to doubt his assertions. Morgan's unwillingness to provide greater rate variety to the initiative have become indicative of a deeper ideological divide than virtually prudent financial management. For Tesla, Wardenclyffe represented a preliminary step in the route of a free-reputation wi-fi electricity future, but Morgan grow to be endorsed thru the possibility of actual, immediate income.

The closing straw have become Morgan's unequivocal rejection of in addition help. Due to the economic network's mistrust and his mounting recognition for broken ensures, Tesla's tries to benefit more capital proved fruitless. The assignment, as quickly as tipped to be a hobby-changer, modified into deserted, a photograph of unrealized promise.

For Tesla, the failure of the Wardenclyffe challenge come to be not just a economic loss but moreover a setback that might in no way in reality heal on a non-public and expert diploma. Once a shining instance of present day technology, Wardenclyffe's enforcing constructing is now a somber mirrored image of what could have been. After a romance based totally on mutual admiration and notable expectancies ended in unhappiness, Tesla positioned himself on my own and alienated within the medical and cutting-edge community.

Wardenclyffe serves as a poignant instance of the troubles that frequently befall trailblazing visionaries: the war to recognise groundbreaking requirements within the face of monetary realities and the commonly careful mind-set closer to making an funding in unproven technology. For Tesla, it have become a time of super promise observed via extreme disappointment; it served as a sobering reminder of the thin line that separates genius from impracticality in the harsh international of industrial agency and medical improvement.

The Downfall of Wardenclyffe

A full-size turning detail in Tesla's existence story, the fall apart of the Wardenclyffe Project represents each the height of his current talents and the lowest element of his monetary and expert fortunes. It changed into also his most ambitious and ultimately catastrophic venture. Designed with the bold intention of permitting wi-fi communique international, Wardenclyffe came to symbolize unmet expectancies and reneged guarantees, illustrating the issues Tesla encountered in bringing his first rate ideas to life.

The Wardenclyffe Tower, located near Shoreham, New York, changed into supposed to feature the point of interest of Tesla's audacious plan for wireless energy and information transmission.It grow to be to be a worldwide telecommunication middle and, greater ambitiously, the fulcrum of Tesla's theories on wi-fi transmission of electrical electricity. Tesla's partnership with financier J.P. Morgan regarded to augur a exquisite future for the undertaking, with Morgan imparting the initial investment based totally on Tesla's promise of a cutting-edge wireless telegraphy device.

But as soon as art work were given underway, Tesla's plans grew a good deal beyond what emerge as first predicted in terms of each generation and fee. The challenge's fees skyrocketed, and Tesla started to invite Morgan for more money greater regularly and urgently. At first know-how and inspiring, Morgan grew extra dubious and careful as Tesla's claims of a feasible tool grew more extravagant and lacked actual-international evidence.

Tesla had more troubles because of the clinical and technological breakthroughs made thru manner of his buddies. Tesla's assertions about the superiority of his private machine have been critically weakened via the usage of using Guglielmo Marconi's 1901 a success transatlantic wireless transmission, which have become finished with an extended manner much less complex and greater low cost system.This fulfillment, coupled with Marconi's backing with the aid of influential financiers and the general public's developing skepticism of Tesla's unproven claims, further eroded aid for Wardenclyffe.

Transatlantic Wireless transmission

Morgan added the closing blow whilst he firmly became off the monetary assist. Without Morgan's assist, different buyers have been hesitant to invest in what emerge as starting to look like a loopy concept. Due to a decline in his price range, Tesla modified into pressured to mortgage the Wardenclyffe property, which he finally presented. Formerly a shining example of current-day generation, the tower now remained as a desolate reminder of a cause left unrealized.

A tangible and symbolic indication of the task's loss of lifestyles became the 1917 destruction of the Wardenclyffe Tower, which become mandated to pay off Tesla's debts.Tesla became disappointed because of the truth he had invested all of his cash, coronary heart, and soul into Wardenclyffe. An essential turning element in his profession and reputation turn out to be the task's failure. Tesla out of place his recognition because the acclaimed visionary of his children and come to be increasingly visible as a "mad scientist" who changed into disconnected from the realities of technological expertise and enterprise.

Even even though it modified right into a devastating financial ruin in Tesla's lifestyles, the Wardenclyffe episode represents his unshakable spirit and unyielding devotion to his thoughts. The venture demonstrates Tesla's brilliance and foresight although it failed. The echoes of Tesla's audacious purpose at Wardenclyffe but reverberate nowadays, reminding us of the high-quality line retaining apart genius and madness in addition to the often-overlooked debt our current global owes to the visionaries of the beyond. We stay in a global connected by using wi-fi era.

Later Years and Challenges

Living in New York Hotels

In sharp evaluation to his in advance years of prolific invention and giant popularity, Nikola Tesla's later years were characterised through a progressive retreat from public life and a growing seclusion. His dwelling in unique inns in New York in the course of this time end up one of the maximum heartbreaking factors of it. He lived from room to room, growing increasingly more reclusive and his economic situation greater insecure.

In the early 20th century, New York's inns served as more than virtually locations to stay; they had been moreover social hubs, architectural wonders, and representations of the metropolis's developing riches and culture. Once a famous inventor and member of New York society, Tesla ended up living in these inns now not due to the truth he preferred to be a high-priced traveler but as an alternative due to the truth he had out of place maximum of his fortune and couldn't locate the money for to keep a eternal house or lab.

Hotel in the 20th

Tesla's lodge living commenced inside the Waldorf Astoria, one of the most high priced houses in New York. In the early 1900s, his rental turn out to be a hub of pastime in which he had meetings with scientists, journalists, and buyers. But while his wealth declined, he turned into needed to relocate to a extraordinary deal much less opulent quarters. He spent the final ten years of his life in The Hotel New Yorker, which turned into his very last domestic. His rooms inside the ones accommodations have been every his sanctuary and his cage – locations in which he have to contemplate and paintings, but furthermore stark reminders of his faded glory and economic misery.

Living in these hotels, Tesla led a solitary lifestyles, his interactions largely confined to the staff and a small circle of pals and benefactors who remained loyal to him. His every day recurring become rigid and whimsical – he might feed pigeons within the park, dine by myself, and preserve his artwork on numerous innovations and theories. Despite his economic struggles, he maintained a dignified look, continuously well-dressed, projecting the picture of a incredible scientist, albeit person who lived inside the shadows of his former self.

Additionally, Tesla's hotel rooms functioned as brief labs. Ever the innovator at heart, he stored working on one-of-a-kind requirements and innovations, making notes and sketches. But those endeavors stayed essentially in the international of idea and fantasy without the investment and backing he previously enjoyed. During this time, Tesla emerge as well-known for his ferocious privateness protection, and there were many unfounded claims regarding the nature of his artwork and his kingdom of thoughts.

His lodge life tells a bigger story about Tesla's later years: a visionary inventor who had become disoriented through way of the practical desires of the company worldwide, and a colorful mind isolated with the aid of its personal knowledge and oddities. These inns were witness to his unyielding spirit, his relentless pursuit of knowledge, and his undiminished want that his improvements may also want to in the end discover the recognition and alertness he believed they deserved.

Over time, Tesla's seclusion have turn out to be increasingly obvious. His public appearances became infrequent, and the social invitations that he had received in the past decreased. Tesla, the man whose innovations as soon as illuminated the globe, lived a appreciably obscure twilight years in those hotel rooms, his modern-day contributions to technological know-how and technology definitely unacknowledged.

Tesla's existence in New York accommodations epitomizes the duality of his being a person who changed the current worldwide however who moreover grew more and more estranged from it. It serves as a shifting reminder of the ephemeral nature of fulfillment and notoriety similarly to the thin line keeping apart brilliance from obscurity.

Public Recognition and Financial Difficulties

Nikola Tesla The Visionary Genius

The dramatic contrast amongst growing public repute and developing monetary troubles characterised Nikola Tesla's later years. Tesla, extended a image of invention, struggled to comply to a manner of life that each preferred and misinterpreted his contributions as the 20 th century went on.

Tesla's in advance groundbreaking achievements served as a catalyst for his public publicity for the duration of this time. He have become praised for his contributions to wi-fi communique and electric powered engineering, and his prediction of an alternating contemporary-powered destiny had come actual.In 1917, he grow to be offered the distinguished Edison Medal thru using the American Institute of Electrical Engineers, a poignant recognition considering his preceding contention with Edison. This accolade, amongst others, served as a testomony to his enduring impact at the area of electrical engineering.

Tesla expert excessive economic troubles in spite of these accolades. A big a part of his profits have been spent on his in advance initiatives, inclusive of the formidable Wardenclyffe Tower, and his propensity to put money into futuristic and from time to time unrealistic thoughts high-quality made his financial problems worse. Moving from inn to inn in a few unspecified time inside the destiny of his life in New York, Tesla often left unpaid invoices in his wake. His way of residing come to be supported via a combination of declining monetary financial savings, royalties from his patents, and the sporadic donations from folks that concept specially of him.

Tesla's demeanor and style of doing employer contributed to his monetary troubles.He have become more of a dreamer than an entrepreneur, regularly greater centered at the quest of knowledge than the ability for benefit from his creations. Though admirable in its quest for scientific improvement, this mindset left him exposed to the difficult realities of organization capitalism. Even even though Tesla had many and revolutionary patents, his royalties have been no longer a steady supply of earnings, and his patents had been frequently involved in courtroom disputes.

Chapter 8: The Grasp Of Lightning

The very last a long term of the inventor's lifestyles were marked by using the use of the use of developing seclusion. He stored operating on new theories and upgrades, together with concepts for wireless strength transmission and novel propulsion structures, however neither the overall public nor the monetary community responded favorably to his efforts. With the generation available at the time, quite some Tesla's creative concepts were now not viable or sensible.

In clinical circles, Tesla maintained his respectability in spite of these problems. He saved getting honors and popularity, and he saved in contact with other engineers and scientists. However, loneliness and unstable rate range have been turning into increasingly capabilities of his lifestyles. His early achievements and next setbacks assessment sharply, illustrating a man whose brilliance have become not always in harmony with the realities of his surroundings.

Consequently, Tesla's latter years have been characterised with the resource of manner of the dichotomy of a person hailed for his in advance accomplishments but locating it tough to set up himself in a society that had outgrown his maximum avant-garde thoughts. His lifestyles sooner or later of this period is a tribute to the troubles encountered thru numerous visionaries: being beforehand of 1's time can bring about a lifestyles filled with miscommunication and adversity in addition to a long-lasting legacy.

Tesla's Eccentricities and Loneliness

Tesla withdrew proper into a worldwide of his very private introduction, one complete of wild mind, unproven improvements, and everyday conduct, as his financial circumstance worsened and precise innovators' contributions to generation grew more and more obvious. Unmarried, Tesla stayed on my own himself in his New York inn rooms, in maximum cases on the New Yorker Hotel. He chose to live by myself, it wasn't truly an problem of situation. His perception within the need of last single for his modern and scientific endeavors led him to make the well-known declaration that he couldn't see engaging in the same diploma of achievement if he had a circle of relatives. But as he grew greater reduce off from the out of doors international in his very last years, this practical seclusion just made him sense greater by myself.

A notably mentioned thing of Tesla's eccentricities changed into his deep affection for pigeons. He took extraordinary care of them and modified into regularly observed feeding them in parks or through the open window of his lodge room. Tesla claimed to like a white pigeon as plenty as a person loves a female, and the 2 had a mainly close to courting. When the pigeon surpassed away, he end up distraught and taken into consideration its demise to be a super non-public tragedy. This strong emotional bond that Tesla had with pigeons modified right into a meditated image of his loneliness and shortage of potential to locate connection amongst his fellow people.

Tesla has a totally prepared and positive each day agenda. He had his food prepared a positive manner and ate on the same consuming places each day on the equal time.His obsession with the quantity 3 come to be another example of his obsessive-compulsive conduct; he need to regularly stroll round a block 3 instances in advance than getting into a constructing and will typically insist on having a stack of 3 folded napkins subsequent to his plate at mealtime. Although the ones rituals gave him a feel of order and control in his life, they seemed unreasonable to outsiders and were a sign of a seriously disturbed thoughts.

Nikolas Tesla and pigeons

Tesla continued to artwork on new theories and improvements in spite of his solitude, however plenty of those later tasks were in no way determined out. He positioned out ever extra ridiculous standards, at the side of a dying ray that would placed a prevent to fight with the beneficial resource of making an impenetrable barrier. The medical community especially unnoticed these theories, seeing them because the musings of a as soon as-notable mind that had end up detached from truth.

Tesla's loneliness become exacerbated with the useful resource of his monetary troubles. He used to live in luxurious, but in his later years, he have end up in big element impoverished, his former prosperity and achievements extended prolonged long gone. He lived in a succession of inn rooms and changed into depending on a minimal stipend supplied thru Westinghouse Electric & Manufacturing Company; his money owed often ended in his eviction. For Tesla, who had dined with the top schooling and counted a number of the maximum famous humans of the day amongst his pals, his economic problems were a motive of fear and humiliation.

As his peculiarities and private problems eclipsed his achievements to era and technology, Tesla come to be often considered in those very last years as a recluse and misunderstood genius. Nevertheless, regardless of his eccentricities and loneliness, Tesla in no way gave up on his vision of an invented future. His continual willpower to investigate and invention is confirmed through using his unflinching religion within the opportunities of his requirements, even within the face of rejection and absence of help. Tesla's peculiarities and loneliness in his later years advise a man who, having altered the path of information, felt himself increasingly at odds with it. In those final chapters, his lifestyles is as a notable deal a story of clinical brilliance as it is of unrealized functionality and human sorrow.

Tesla's Last Years

Continued Inventions and Theories

The very last years of Nikola Tesla's existence had been marked via the usage of each extreme inventiveness and heartbreaking warfare. Tesla's thoughts was constantly churning out thoughts, from the modern to the fantastical, no matter his rate variety and public attention deteriorating and him being a man of many inventions and theories. His non-save you devotion to innovation at some point of this degree of his life is proof of his steadfast religion in the electricity of human creativity and his unflinching devotion to technology.

As his economic assets faded, Tesla spent those years residing in some of motels in New York, in ever-more-modest quarters. He frequently used his resort room as a workshop, where he had a mess of papers, diagrams, and tiny experimental devices. Tesla's creativity have end up unrestricted with the aid of the confined environment. He experimented with many exceptional areas, bobbing up with mind for emblem spanking new kinds of dynamos, blade-much less generators, and wi-fi transmission that went past his preceding conceptions.

During his very last years, Tesla's maximum bold reason changed into to gather a particle beam weapon, a concept he known as "Teleforce." With the ability to down enemy aircraft and so prevent struggle, he noticed this system as a technique of shielding in opposition to aerial attacks. According to Tesla, this innovation must produce a sturdy beam of strength via combining mechanical and electric powered strategies. Despite his excellent efforts, the notion by no means have become a actual enterprise, no matter the fact that he pitched it to severa nations, which encompass the usa, UK, and USSR.

Nikolas Teleforce called "Death Ray

Tesla moreover dabbled in physics and cosmology together along with his notions. In evaluation to the then-dominant clinical theories, he proposed the existence of a dynamic ether that pervaded vicinity in his theories about the man or woman of the universe. Additionally, he stored refining his theories on wi-fi energy transmission, thinking that it would be viable to apply the strength contained in Earth's surroundings to distribute energy inside the course of the entire planet.

Tesla displayed proof of obsessive-compulsive behavior and unique oddities during his scientific pastimes. He advanced a sturdy dislike for specific office work and jewelry, accompanied a strict every day time table, and professed to keep up a correspondence with extraterrestrial entities. Notwithstanding the ones eccentricities, Tesla maintained his readability in his clinical reasoning and showed that he understood the underlying ideas and ramifications of his discoveries.

It become wonderful how resilient Tesla remained in the face of trouble. He stored up his article writing, interviewing, and inventing paintings at the same time as his physical fitness deteriorated and his monetary popularity grew more risky.Throughout the ones conversations, he regularly stated his lofty future desires, which covered the development of generation for interplanetary communique, the chance of wirelessly doling out energy over great distances, and the capability for solar and wind energy.

When one considers Tesla's latter years, it becomes smooth that his ambition have end up an lousy lot past the kingdom of era on the time. Many of his mind in no way got here to bypass in his lifetime, and some have been so present day for the time that they may be even though unexplored nowadays. But this come to be not only a tale of desires unmet and fortunes falling from grace at some point of this time in his life. It served as proof of Tesla's steadfast hobby and his deep faith inside the capacity of clinical development to exchange the path of data.Tesla's final years, marked with the resource of relentless creativity in the face of overwhelming demanding situations, characteristic a poignant reminder of the iconic spirit of without a doubt one of statistics's high-quality inventors.

Chapter 9: Death And The Aftermath

The closing years of Nikola Tesla had been characterised thru way of a moving combination of economic issue, intellectual tenacity, and seclusion. Tesla maintained his enthusiasm for creation and discovery in his artwork whilst his public profile faded Living at a string of New York motels, generally the New Yorker, Tesla's existence modified into a miles cry from his extra famous and celebrated days. His financially volatile situation resulted from the expiration of his once beneficial patents and the decline in the quantity of monetary assist from benefactors.

Tesla's physical and highbrow properly-being had began out to end up worse. His ordinary behavior and isolated life-style made his various maladies worse. He was tormented by a growing range of ailments. With all of this, Tesla endured in his determination to his activity, arising with new mind and filing patent packages. He took great satisfaction in tending to the pigeons inside the municipal parks, a hobby that gave him a few consolation in his latter years.

Nikola Tesla died on my own in his New Yorker Hotel room on January 7, 1943. Coronary thrombosis become located to be the purpose of dying. Tesla grow to be 80-six. With his passing, a duration came to an cease and a existence dedicated to the boom of generation and generation got here to an quit. The press gave the statistics of his loss of lifestyles scant insurance, a depressing indication of the manner out of the general public's interest Tesla had slipped.

Following his passing, america government confiscated Tesla's non-public assets, research substances, and documents. Fearing that his artwork might likely have navy programs, the Office of Alien Property seized his possessions. The nature and significance of Tesla's unpublished art work had been the challenge of many thoughts and conjectures, specially with regards to his experiments with wi-fi power transmission and one of a kind floor-breaking technologies.

Thousands of human beings, which encompass many top notch web page traffic and participants of the medical community, attended Tesla's funeral, that have come to be hung on the Cathedral of Saint John the Divine in New York City.His stays have been incinerated, and his ashes were deposited in a golden urn usual like a sphere, signifying his audacious pursuit of global wireless strength transmission. The urn, a becoming memorial to a person whose ancestry have end up as international as his aspirations, is on show off on the Nikola Tesla Museum in Belgrade, Serbia.

Following his passing, there has been a slow but massive resurgence of interest in Tesla's life and contributions. As the value of Tesla's contributions to modern science and technology extended over the years, he modified into honored posthumously as a trailblazing inventor and a visionary ahead of his time. Worldwide, there are without a doubt monuments, museums, and memorials erected in his honor, and his call is now associated with originality and inventiveness.

In the contemporary-day technology, Tesla's imaginative and prescient of a international powered by means of manner of renewable electricity and related through wireless era is turning into an increasing number of crucial. His theories, which were before everything appeared as radical and impractical, are resonating in a society that is looking for technical answers which is probably related and sustainable. In addition to the technology he contributed to boom, Tesla left within the returned of an progressive and insatiably curious mentality that endures nowadays. His lifestyles serves for example of the innovative potential and the unwavering pursuit of statistics, encouraging limitless numbers of scientists, engineers, and dreamers to appearance past the restrictions of the proper here and now and take into account an infinitely bright destiny.

Tesla's Legacy

The Revival of Interest in Tesla

As fascinating as the inventor himself is the phenomena of Nikola Tesla's rebirth in recent a few years. After being driven to the outer fringe of data and eclipsed by using his friends, Tesla has come over again to lifestyles as a charming determine who has almost attracted a cult following. He personifies the stereotype of the misunderstood genius whose thoughts had been some distance beforehand in their time.

There are numerous motives for this resurgence. First off, the internet and the virtual age have been crucial in locating and sharing information approximately Tesla's life and contributions. An worldwide audience can now access a multitude of materials, consisting of as his books, patents, and lecture transcripts, which has introduced about a renewed appreciation of his contributions to technological know-how and generation.

A sparkling angle on Tesla's contributions has also been provided through way of the environmental movement and the hunt for sustainable energy alternatives. His early research on wi-fi energy transmission and his guide of the use of renewable belongings have a widespread effect on a international that is struggling with climate alternate and the depletion of fossil fuels. In addition to being groundbreaking, Tesla's idea of a easy, loose electricity-powered future moreover seems quite pertinent these days.

As a lone genius suffering in opposition to a civilization that did not apprehend him, Tesla has been provided as a sad hero in severa movies, books, and television series which have immortalized him in well-known tradition. While it's far proper that some elements of his existence were idealized on this example, it has moreover contributed to the overall public's awareness of his tale.

The 21st century's generation innovations—mainly in regions like wireless communique and electric powered powered vehicles—have highlighted how in advance in their time Tesla's thoughts have been. The most famous instance of his continuing have an effect on is probable the electrical automobile producer Tesla Motors, which bears his call. It represents every the fulfillment of Tesla's aim of electrical mobility and the broader software program program of his energy and innovation concepts.

Nowadays, there can be more acknowledgement and birthday celebration of

Tesla's contribution to the improvement of alternating modern-day (AC) power, which runs our towns and homes. His triumph inside the War of Currents is now visible as a watershed within the development of electrical engineering, having formerly been not noted in select out of Edison's legacy.

There have been a ton of biographies, instructional papers, and meetings dedicated to reevaluating Tesla's paintings, reflecting the extended hobby in his work among teachers. With this scholarly cognizance, severa of the misconceptions surrounding Tesla had been debunked, giving rise to a more nuanced portrait of his accomplishments and troubles.

Additionally, Tesla has come to represent the bigger story of the unsung hero in technology and technology. His narrative activates a reexamination of records intending to widely diagnosed numerous and unconventional contributions, and it serves as a reminder of the severa pioneers whose contributions have been left out or underappreciated.

Essentially, the resurgence of interest in Nikola Tesla is not pleasant a ancient reevaluation; as an opportunity, it is a scientific and cultural awakening to the legacy of someone who imagined generation transforming the sector. Once out of vicinity to data, his lifestyles's art work is now an concept to a extremely-cutting-edge era of engineers, scientists, and dreamers. Through rediscovering Tesla, we decorate interest of the potential for human ingenuity on the same time as moreover honoring his genius.

A portrait of Nikolas Tesla's Structure Plan

Chapter 10: Impact On Modern Science And Technology

In the world of present day technological know-how and generation, Nikola Tesla left in the back of a considerable and sundry legacy. His thoughts and creations installed the inspiration for many innovations that characterize contemporary-day-day dwelling. Perhaps most significantly, the muse of the modern-day electrical grid is Tesla's groundbreaking paintings in alternating present day (AC) energy systems. The AC machine transformed the way strength become transmitted globally, transferring from localized electricity technology to large-scale, related strength networks. This modified into made viable with the aid of using its efficient ability to hold electricity over superb distances. This alternate ushered in a period of rapid commercial enterprise improvement and clinical soar forward, permitting the electrification of cities and the widespread use of electrical merchandise that basically altered every day existence.

Tesla's discoveries and theoretical paintings have had an extended-lasting impact on a number of sectors in addition to the AC system. His invention of the AC-powered induction motor provided a greater dependable and inexperienced alternative for the direct modern-day-day (DC) motors which have been in use at the time. This invention become extensively implemented in industry to power the whole lot from contemporary electric powered powered powered automobiles to production unit device. Additionally, Tesla's research of electromagnetic fields and excessive-frequency currents spread out new tips for wi-fi conversation. Even even though it changed into in no manner absolutely discovered out within the path of his lifetime, his concept of wi-fi electricity transmission stimulated later enhancements in radio technology. His thoughts offer the concept of all modern-day wireless communication systems, which include radio, tv, and the cell networks that assist the arena's infrastructure of communications.

Concurrent with Wilhelm Röntgen's discovery, Tesla's paintings with X-rays proved his imaginative and prescient inside the realm of medical imaging. Even at the same time as his contributions to this place acquired a good deal less popularity, they however confirmed that he changed into capable of increase his highbrow hobby out of doors the sector of electrical engineering. In a comparable vein, his work with radio-controlled devices foreshadowed the improvement of robotics and some distance flung manipulate generation, which is probably now critical to a extensive sort of sectors, from production to place exploration.

Wilhelm Röntgen

Tesla made many theoretical claims which have been a top notch deal in advance in their time, specially within the regions of renewable energy and wireless energy transfer. His concept of using the energy of natural occurrences, together with wind and sun power, can be very similar to present day tasks to create sustainable power assets.Tesla's theories foresaw in lots of respects the present fashion in the direction of renewable energy and the non-prevent development of structures to harness those assets.

Furthermore, the interconnected global of the twenty-first century changed into foreseen thru Tesla in his thought of a "global wi-fi gadget" for the transmission of power and statistics. The spirit of Tesla's formidable purpose is embodied by brand new global internet and satellite television for laptop television for pc communication networks, regardless of the truth that the era of his day turned into now not capable of perform this vision. His high-quality capability to envisage advances that might no longer be located out till many years after his loss of life is established with the resource of his vision of a society in which strength and facts can be conveyed wirelessly.

Beyond precise upgrades and theoretical contributions, Tesla has stimulated innovation and scientific interest collectively along with his approach to technological know-how. For upcoming generations of scientists and inventors, his emphasis on creativity, meticulous experimentation, and a comprehensive comprehension of natural techniques installed the equal vintage. His struggles to decorate unconventional theories and address the realities of commercializing discoveries provide valuable insights into the connection amongst clinical discovery, entrepreneurship, and societal influence.

Tesla in Popular Culture

Nikola Tesla is a timeless parent in well-known way of life, representing the archetypal misunderstood genius whose thoughts had been earlier of his time. His impact is going well past the fields of generation and generation. His lifestyles narrative has captivated the hobby of artists, writers, filmmakers, and the general public due to the reality that it's miles filled with extremely good accomplishments and heartbreaking struggles. This cultural interest is a meditated picture of each intrigue alongside collectively with his mysterious nature and ahead-thinking ideas, further to an appreciation of his scientific contributions.

Tesla has a extensive-ranging have an effect on on famous tradition. He has been represented as a clever but eccentric inventor in a number of movies, books, and television programs. Tesla has constantly been a mysterious and enigmatic discern in movies, introduced to life through famend performers who capture his distinct personality and sharp mind. These depictions, which capture the dramatic tale of his life, frequently highlight his opposition with Thomas Edison, his financial hardships, and his floor-breaking improvements.

Tesla has been the hassle of fiction similarly to non-fiction in literature, which has brought on the emergence of a subgenre of works that combine ancient records with fantastical fiction. His life's dramatic trajectory—from his early accomplishments to his later years of poverty and obscurity—has attracted the attention of writers. Stories of technology, thriller, and creativity abound manner to Tesla's imaginative mind and innovations. In the ones works, his character is often provided as a illustration of the unwavering quest of information and creativity in the face of social and now not pricey boundaries.

Tesla's cultural have an effect on has multiplied a good deal more inside the age of the net. Social media web websites and on line boards have contributed to a renewed interest in his lifestyles and art work. On the internet, Tesla has acquired reputation and is frequently hailed as an inventor and a era hero. There are many web sites, net forums, and social media pages devoted to Tesla, wherein humans debate his recognition some of the excellent scientists and inventors of all time, look at his creations, and speculate about his unfulfilled thoughts.

Bands and musicians have taken concept from Tesla's lifestyles and art work, demonstrating how an prolonged way his impact reaches into the humanities. His story has been weaved into track that resonates with problems of invention, tenacity, and the pursuit of understanding. Songs and albums have been named after him.

Tesla's have an impact on may be seen in some of ingenious mediums, such as virtual and conventional art. His incredible talents, the dramatic visuals of his experiments, and the metaphorical importance of his creations have all captured the eye of artists. Tesla is often portrayed in paintings that evokes awe, mystery, and the ethereal nice of his medical endeavors.

One of the most vital signs and signs and symptoms of Tesla's lasting affect in famous lifestyle is probably the relationship his call has with innovation and current-day technology. The Tesla electric powered automobile, which bears his call, represents a destiny in which current-day era and renewable strength resources propel human improvement. This memorial to Tesla is proof of his enduring have an effect on and the imaginative and prescient's ongoing importance.

Chapter 11: The Genius Innovator Ahead Of His Time

You probable do now not pay a whole lot idea to how a few detail works or to the folks who made all of it possible whilst you turn on a transfer or a lamp. You might possibly say Thomas Alva Edison, the person who created the incandescent moderate bulb in case you had been required to find out the genius behind the lamp. Yet, a visionary with the resource of the call of Nikola Tesla changed into equally as influential, if no longer extra so.

One of the best inventors and visionaries in facts, Nikola Tesla is thought for his ingenuity and genius. Tesla, who turned into born on July 10th, 1856, in Smiljan, Croatia, changed electrical engineering and the manner we harness and use electricity The basis for Tesla's first-rate profession have come to be constructed thru his upbringing and education. He began working for famed inventor Thomas Edison after completing his engineering and physics university. Yet, their divergent views on the development of energy precipitated a sour war called the "War of Currents". From a younger age, Tesla confirmed a tremendous flair for technological statistics and engineering. He changed into inquisitive about the herbal international, and he spent hours experimenting with electric circuits and extraordinary medical phenomena. He changed into also a voracious reader, and he devoured books on physics, mathematics, and distinct clinical subjects. At an early age, Tesla's intelligence and creativity were smooth to see. He have come to be able to mentally see elaborate systems and techniques, and he ought to provide you with modern-day answers to troubles that appeared no longer viable to solve. He emerge as a expert artist as properly, and his medical artwork frequently contemplated his aesthetic sensibility.

Tesla's life modified into no longer with out its problems, irrespective of his severa accomplishments. The scientific and engineering popularity quo of his time frequently misunderstood and undervalued him, and he struggled to acquire financial manual for his studies. Additionally, he struggled with intellectual fitness problems inclusive of hysteria and despair, making it frequently tough for him to maintain strong relationships. Tesla was a complicated and multifaceted character, and his person and creativity hold to inspire new generations of scientists and engineers. He became a first rate inventor and engineer; however, he become moreover a deeply compassionate and empathetic character who become dedicated to the use of his skills to make the place a higher vicinity.

Tesla is thought for his eccentric persona and unusual conduct. He has an brilliant reminiscence and claims to have visualized his innovations vividly in advance than making them. Tesla moreover come to be fascinated by pigeons and had a strong reference to them, even claiming to acquire crucial records and concept from this fowl. Despite his genius, Tesla faced economic problems within the route of his lifestyles and struggled to get the popularity he deserved. He frequently prioritizes his artwork over monetary gain, leaving a number of his projects unfinished and inventions unpatented. However, Tesla's effect on cutting-edge-day society can't be overstated. His improvements laid the principles for the electrification of the arena and paved the way for loads technological advances. Tesla died on January 7, 1943, leaving in the back of a legacy that continues to encourage scientists, engineers, and inventors. His imaginative and prescient and foresight paved the way for the development of the myriad technologies we rely on nowadays, from the power grid and wireless communications to electric powered automobiles and renewable energy systems. As we experience the stop stop end result of his labour, it's miles vital to understand and appreciate the profound impact Tesla had on shaping the destiny of electricity and technology.

PICTURING, DESIGNING, INVENTING

The legendary electric powered engineering genius and inventor Nikola Tesla had a mind that was past the regular. Unique modern visualization, a photographic reminiscence, and strong practicality set him other than his buddies and have been essential to his modern approach. A testament to his exceptional cognitive skills, Tesla emerge as capable of photograph his improvements with astonishment, meticulously format them, and then deliver them to life. Tesla's great capacity to visualise changed into on the coronary coronary heart of his progressive machine. He must paint complex and unique pics of his enhancements on a canvas like his thoughts did. Tesla have to immerse himself in huge intellectual exploration in advance than starting any form of bodily introduction. With eyes shut, he must intellectually acquire and control the particular components, imagining their precise shape, capability, and interconnections. Tesla's visualisation technique end up a colourful and powerful experience in choice to genuinely a shape of having a pipe dream. He need to view his creations from pretty some angles and views due to the pretty precise snap shots that might seem in his thoughts's eye. He had the capability to mentally hone and perfect plans, confirming their effectiveness and viability earlier than moving ahead with the physical execution. Tesla's photographic reminiscence completed a crucial role in his excellent functionality for visualization. He had a notable ability to take in a splendid quantity of facts, from mathematical calculations and clinical theories to mechanical and electric powered engineering thoughts. Tesla changed into able to keep in mind and preserve a outstanding quantity of records efficaciously because of the reality his mind worked like a big library. His innovative thoughts had been formed on the foundation of this intellectual database of know-how. His private theoretical theories and observations made whilst doing his exams led him to be sceptical of the typically common model of atomic shape. Tesla disagreed with the conventional medical cognizance that publicizes that electrons and distinct tiny subatomic debris make up an atom. He stated that the idea of electrons producing an electric powered charge come to be wrong and that electrons, if any existed at all, had no direct connection to energy. Tesla proposed a fourth kingdom of consider, which he referred to as an "virtual united states" or "sub-atom". In assessment to everyday atmospheric situations, he hypothesized that this nation of count wide variety could nice be determined in a really rarefied or experimental vacuum. Tesla held the belief that this sub-atom nation changed into the reason of electrical phenomena and had exceptional homes. Nikola Tesla's claims about his very personal physical precept and his alleged improvement of a dynamic idea of gravity have been the hassle of a whole lot of hypothesis and debate. While Tesla made statements related to those thoughts, it is vital to word that he did no longer supply comprehensive written documentation or medical papers outlining those theories. As a end end result, the unique nature and validity of Tesla's claims continue to be unsure, and no, addition, proof or elaboration of these theories has been observed in his writings. Without any helping proof or posted works through Tesla, it is hard to assess the specifics of his assertions of a bodily precept and a dynamic idea of gravity. Although it's far feasible that Tesla changed into experimenting with novel requirements and theories concerning take into account huge variety, electricity, and gravity, it is hard to evaluate the clinical validity or awesome of those statements inside the absence of thorough justifications or assisting facts.

Tesla's photographic reminiscence enabled him to draw on beyond reviews, and observations in building his highbrow photographs. He want to without troubles don't forget clinical literature, diagrams, and equations and seamlessly integrate them into his highbrow visualizations. This precise aggregate of visible memory and encyclopaedic facts of Tesla lets in the layout of complicated and technically sound innovations. What prominent Tesla modified into his practical depth in the revolutionary approach. His visualizations had been not imaginitive; they were carefully designed with practicality and real-international software program in mind. Tesla emerge as aware of the limitations of present generation and the realistic demanding situations they could face. Therefore, his highbrow designs were no longer mere fantasies, but modern solutions that took under interest technical barriers and feasibility. Tesla's practicality became established in his ability to anticipate capability issues and create current-day answers internal his intellectual visualizations. He mentally ran simulations and assessments, thinking about severa situations and refining his designs to mitigate potential flaws or imperfections. This pragmatic method, combined along collectively along with his incredible visualization skills, allowed Tesla to introduce progressive improvements that modified the arena. Furthermore, Tesla's pictures were no longer limited to single improvements; he may additionally moreover want to appearance the interconnectedness of his works within a larger system. He had a unique potential to expect the aggregate of more than one innovations and technology and the manner they could work in synergy and create transformative effect.

In his autobiography, Nikola Tesla recounts his first-rate ability to visualise a selected equipment and eventually take a look at-run it in his thoughts, disassemble it mentally, and compare its overall performance and scenario. This exquisite gadget highlights Tesla's excellent cognitive skills and his profound knowledge of engineering requirements. After trying out the system mentally, Tesla have to keep to disassemble it in his thoughts's eye. This worried mentally taking apart the additives, studying them, and assessing their put on and tear. Through his top notch visible go through in mind and complete information of engineering ideas, he want to study the condition of every element and decide if any adjustments or replacements were essential. Tesla's particularly evolved creativeness and fantastic spatial awareness allowed him to envisage a complicated tool with such clarity and precision. He had a tremendous potential to visualise complex designs and precisely positioned them collectively in his thoughts. He have emerge as able to see the device's severa factors, their placement, and their linkages thru using imagining it. Tesla might imagine the device being built and then simulate how it'd art work. This comprised jogging the tool in a simulated placing, watching the way it behaved, and comparing how properly it finished. He changed into able to spot capability weaknesses, inefficiencies, or areas that wished similarly paintings through this intellectual simulation. The path of intellectually dismantling the mechanical meeting time-honored Tesla to acquire bits of information approximately not unusual revel in and the unwavering tremendous of his plan. He ought to pick out capacity prone factors or failure-susceptible areas by using way of evaluating the components' placed on and tear. He have emerge as guided via this data as he sensitive the format and made the necessary adjustments to enhance the system's average performance and durability. It is important to keep in thoughts that Tesla's top notch visualization capabilities have been now not the best detail in his potential to mentally take a look at, run, and disassemble an tool. His in-depth comprehension of the underlying clinical and engineering necessities that ruled the system's behaviour additionally contributed to it. Tesla's huge facts and involvement with electrical designing permitted him to precisely re-enact the connections amongst components and foresee their usefulness. The accuracy of Tesla's highbrow attempting out and disassembling technique have end up often proven whilst he bodily built and tested the device in fact. More frequently than not, the actual normal performance became cautiously aligned alongside along with his highbrow simulations, attesting to the reliability of his visualization competencies. His inventive visualization, coupled together along with his huge expertise and sensible revel in, allowed him to simulate the behaviour and don't forget the scenario of the machine with exquisite accuracy. Tesla's tremendous technique demonstrates the electricity of the human concept while harnessed to its fullest capability.

As we said in advance, Tesla possessed a photographic memory, a mental school that allowed him to preserve one-of-a-type visible facts with terrific accuracy. This supposed that he should memorize problematic blueprints and specifications of his inventions in truth through analyzing them. Once he internalized this records, he no longer had to talk over with bodily blueprints or written documents at a few stage inside the generating phase. Having the blueprints and specs ingrained in his reminiscence furnished Tesla with numerous blessings. Firstly, it allowed him to paintings with super performance. Without the want to continuously are searching for advice from physical files, he may additionally additionally want to continue with the manufacturing technique seamlessly, disposing of time-ingesting steps which consist of retrieving, reading, and interpreting blueprints. This streamlined technique enabled him to artwork suddenly and restrict interruptions, because of this growing productivity. Furthermore, Tesla's capability to work with all the blueprints and specs in his head are extra appropriate for his adaptability and versatility all through the manufacturing manner. In case modifications or changes had been required, he need to shortly recall the critical crucial elements and make the critical modifications except the need for outside references. This saved time and allowed for seamless modifications to the diagram or production manner, ensuring the most effects. Tesla modified into able to create matters without actual blueprints in aspect due to his profound mastery of engineering thoughts. He knew the whole thing there was to apprehend about electric engineering, mathematics, and physics. He modified into capable of recognise the critical ideas and connections within the again of his creations way to this thorough information. Using this information, Tesla may want to mentally photo the development technique, foresee troubles, and make judgments with out continuously concerning drawings.

It should be mentioned that Tesla's potential to work with all of the plans and specifications in his head changed into not restrained to virtually photographic reminiscence. It was the quit give up end result of his relentless pursuit of expertise, relentless curiosity, and consistent quest to recognize the intricacies of his inventions. Tesla's passion for his paintings enabled him to internalize complex requirements and visualize them with wonderful clarity. This enabled him to streamline the manufacturing system, work with the awesome efficiency, and make corrections or modifications on the fly.

Chapter 12: What Tesla Gave The World?

We have heard this genius, Tesla, modified the way we use energy or electricity today. In January 1880, Thomas Edison introduced his electric powered slight bulb to the public, a milestone inside the improvement of electric lights. This seminal invention provided a practical and efficient possibility to gas lamps and paved the way for the huge use of electrical lighting fixtures structures. Shortly after the lamp's advent, Edison began out implementing the present day energy device in New York's First Circuit. Edison's electric system end up designed to permit the big-scale manufacturing, distribution, and use of power. It consisted of severa key additives consisting of generators, transmission traces, and distribution grids. The 1st District of New York served as a trying out ground for this modern device. The set up of Edison's electric powered powered tool in the First Ward delivered electric powered powered lights to the streets, groups, and homes of New York. The previously dimly lit streets had been now illuminated with the brilliance of electrical lamps, improving visibility and safety in the direction of night time time-time hours. The installation additionally allowed companies and homes to enjoy the advantage and usual performance of electrical lighting, converting the greater cumbersome and dangerous gas lighting fixtures structures of the day.

Edison's energy machine changed into primarily based on direct current (DC), which worried the essential generation of strength in energy plant life and its distribution to clients via a cable community. The First Ward of New York City have grow to be a show off for electric powered powered power's capability, demonstrating its feasibility and reliability in sensible applications. The a success implementation of the system in New York City laid the muse for growing energy infrastructure to a similarly vicinity, ultimately leading to the electrification of whole cities and areas. However, Edison's direct contemporary tool had limitations, specially in phrases of transmitting strength over prolonged distances. This added approximately the upward push of Nikola Tesla's alternating cutting-edge (AC) tool, which furnished full-size benefits in phrases of efficiency and extended-distance transmission skills.

The "War of Currents", wherein AC emerged due to the fact the dominant strength transmission generation, could in the end take vicinity due to the competition among Tesla's AC system and Edison's DC device. Edison's advent of the electrical incandescent lamp and the installation of his energy machine inside the First District of New York City had been essential in popularizing electric powered lights and laying the basis for the enormous use of electricity in a number of packages, no matter the eventual occurrence of AC strength systems. The development of realistic electric powered powered powered energy systems and Edison's contributions to electric engineering are though sizable junctures within the data of electrification and characteristic had a long-lasting effect on the way our towns are illuminated and powered today.

Nikola Tesla, the wonderful inventor, and visionary, continuously modified the course of electrical electricity transmission alongside along with his sparkling invention of alternating contemporary-day (AC). Tesla's AC contemporary device revolutionized the way electrical energy is generated, transmitted, and carried out, laying the idea for the present-day electrical grid. Prior to Tesla's AC present-day gadget, the foremost shape of electrical electricity transmission changed into based totally absolutely mostly on direct modern (DC). However, DC strength confronted numerous boundaries, in particular in terms of its functionality to efficaciously transmit electric strength over extended distances. DC structures suffered from huge electricity loss and required well-known and luxurious infrastructure installations to keep voltage degrees. Recognizing the shortcomings of DC, Tesla had been given right down to boom a desire that could conquer those demanding situations. He targeted his efforts on alternating current, a gadget that involved abruptly converting the route of the electrical current. Tesla understood that AC had the attainable to revolutionize electric powered strength transmission because of its capability to transmit electric powered powered strength successfully over prolonged distances with minimal electricity loss.

The invention of the induction motor modified into one of the most essential additives of Tesla's AC device. This motor need to convert electric energy into mechanical energy with the useful aid of using AC energy to create a rotating magnetic issue. Tesla launched into a chain of bold experiments and public demonstrations to illustrate the capability of AC strength. The "War of the Currents", a fierce opposition among Tesla's AC system and Thomas Edison's DC system, is one in every of records's maximum well-known activities. The superiority of Tesla's AC tool changed into proven through powering homes, businesses, or maybe road lighting fixtures over lengthy distances. After the successful implementation of the Niagara Falls hydroelectric energy assignment, Tesla's AC modern-day tool acquired recognition and giant recognition. The practicality and effectiveness of Tesla's invention had been mounted via harnessing Niagara Falls to generate AC power on a large scale. The task signalled the sunrise of a trendy technology in the transmission of power and established the large functionality of AC energy for metropolis areas. Tesla's discovery of AC modern has had an on-going impact. The cutting-edge electric grid, which factors strength to our houses, companies, industries, and present day generation, is largely dependent on the transmission of AC electricity. The worldwide's electrification became made possible through the use of Tesla's innovation, which paved the way for the massive use of electrical domestic device, lights structures, cars, and hundreds different electric powered powered device.

The ground-breaking artwork that Tesla did with a long way off manage changed many industries and set the degree for many new dispositions in automation, conversation, and enjoyment. Not great did he make the a long way flung control more reachable, however he moreover unfolded an entire new international of opportunities which have due to the truth customary our contemporary society. Tesla's interest in a ways off manipulate stemmed from his desire to overcome physical obstacles and permit some distance flung tool control. His innovative thoughts expected a world in which machines can be managed from a distance, liberating human beings from the constraints of bodily contact. This belief, combined collectively together with his profound comprehension of electrical format and faraway transmission, moved Tesla to investigate and foster controller innovation. The development of the tele-automation gadget is one of the tremendous contributions that Tesla made to the sector of remote control. Through the transmission of electrical signals, this tool made it feasible to manipulate mechanical gadgets. By remotely controlling a deliver, Tesla established the functionality of his tele-automation device to govern gadget with out direct human intervention. The value of Tesla's invention of a long way flung manipulation lies in its transformative have an effect on on some industries. One of the earliest and most extremely good abilities changed into within the area of transportation. Tesla's an extended way flung manage technological information paved the manner for the improvement of a ways off-operated motors and guided systems. Today, a ways-off manipulation is notably utilized in the car, aerospace, and maritime sectors, allowing more secure operations, surroundings-exceptional navigation, and more precision. Furthermore, an extended way-flung manipulation era has finished a pivotal feature inside the problem of automation. Tesla's invention laid the basis for the improvement of automatic structures that might want to be managed remotely. This has brought about stepped forward typical overall performance, elevated productiveness, and decreased human intervention in hundreds of industries, starting from production and robotics to agriculture and healthcare. Remote management has converted the way we've got interplay with machines, permitting complicated operations and a long way-off tracking in real-time. The impact of Tesla's invention of the a long way off manage is also obvious in the field of communications. Remote manage generation has been blanketed into severa communication devices, including TVs, radios, and smartphones. This allows customers to effortlessly use and navigate through numerous competencies and talents, improving person experience and comfort. Remote controls have grow to be an vital a part of our each day lives, allowing us to interact remotely with a big wide variety of gadgets.

In addition to sensible applications, the invention of the Tesla a long way off manipulate moreover unfolded new possibilities within the vicinity of entertainment. Remote control toys, drones, and online game controllers have end up well-known, supplying customers with immersive and interactive opinions. The functionality to manipulate gadgets wirelessly has stepped forward enjoyment alternatives, permitting individuals to enter digital worlds, revel in exciting adventures, and discover new horizons. The importance of Tesla's invention of the far flung manipulate extends past tangible programs. It represents a profound shift in the human-system interface, highlighting the power of era to transport beyond physical obstacles. Remote manipulate generation has empowered us to govern and interact with our environment from a distance, fostering comfort, overall performance, and innovation.

Tesla placed the concept of wireless strength transmission to be fascinating and concept that it could be carried out by way of using high-voltage electric discharges. In the past due 1990s, he started experimenting with one-of-a-type wireless electricity transmitter designs. In 1891, he filed a patent for his first invention, a "approach of conveying electric powered energy with out wires".

Tesla coils are the give up result of years of sorting out and perfecting Tesla's actual layout. It includes a number one coil, related to the electricity supply, and a secondary coil, related to a capacitor. When the number one coil is energized, it generates a immoderate-voltage discharge this is transmitted to the secondary coil, in which it's miles amplified and replayed as a excessive-frequency electromagnetic wave. Tesla have become so inspired with the ability of the Tesla coil that he gave numerous public demonstrations of its abilties, which incorporates a famous one at the 1893 World's Fair in Chicago, in which he lit moderate 2 hundred bulbs wirelessly using a Tesla coil. The Tesla loop had severa beneficial programs, which includes a long way flung telecommunication and early TV broadcasting. Vacuum tubes and strong-country transistors, as an alternative, subsequently took their region as more superior electrical transmission strategies have turn out to be to be had.

The traditional steam turbines, which have been the most essential method of manufacturing electric strength on the time, have been no longer as inexperienced or reliable because the Tesla turbine, which end up created to update them. Tesla felt that numerous electricity assets, consisting of water, wind, and geothermal energy, is probably harnessed to generate power the usage of the Tesla turbine. Tesla's format for the Tesla Turbine is based definitely totally on the standards of electromagnetic induction, that is the technique of producing an electric powered powered current-day-day with the aid of transferring a conductor through a magnetic challenge. The turbine includes a rotor surrounded thru a series of constant electromagnetic coils, powered by way of an outdoor electric powered electricity source. As the rotor rotates, it creates a magnetic area that interacts with the table certain coils, inflicting them to generate currents. This current is then fed right into a generator, changing it into usable power.

There are severa realistic makes use of for the Tesla turbine in cutting-edge-day international. It is regularly carried out in business settings to generate energy from nuclear electricity, renewable electricity, fossil fuels, and certainly one of a kind assets. It's appreciably applied to electricity electric powered powered automobiles and offer backup strength for crucial infrastructure, among numerous matters. The Tesla turbine's immoderate overall overall performance and dependability are key blessings. It can generate energy from a large fashion of belongings, which includes wind and geothermal power, which are not usually available or dependable. As a stop result, it is a essential instrument for reducing our reliance on fossil fuels and raising using renewable electricity assets. It is still a important technology in recent times, and new generations of engineers and inventors are normally inspired via its legacy.

Tesla revolutionized wireless energy switch together along with his efforts. In his idealised society, energy lines might not be essential for the wi-fi transfer of strength, which may permit for its effective transmission for the duration of outstanding distances. Although he did now not stay to look his vision of giant wi-fi energy transmission without a doubt realised, his ground-breaking paintings paved the manner for destiny enhancements in wi-fi charging and electricity transfer era.

Chapter 13: A Closer Look At Mr. Tesla!

Several of Nikola Tesla's upgrades went neglected, even as others were misplaced while the fireside burned his notes. His studies became seized with the aid of the FBI at the save you of his existence, and it has surely in recent times been made available to the general public. The Wardenclyffe Tower venture stands as one in each of Nikola Tesla's most formidable and captivating ventures. Conceived as a grand task to transmit wireless strength and revolutionize international communication, the task encountered numerous demanding situations and, in the long run, did not collect its meant objectives. The Wardenclyffe Tower is despite the fact that a tribute to Tesla's formidable necessities and his unwavering pursuit of floor-breaking breakthroughs however its failure. The idea in the back of the Wardenclyffe Tower end up to create a large wireless transmission station capable of transmitting electric powered electricity and communication alerts with out the want for conventional wires or cables. Tesla expected a global in which energy may be harnessed and allotted wirelessly, presenting apparently infinite energy to homes, agencies, and industries across tremendous distances. The tower, over 100 eighty toes tall, will serve as the essential hub for the wireless transmission device. To finance the undertaking, Tesla sought monetary aid from an entire lot of belongings, together with Notable buyers like JPMorgan. However, securing sufficient funding proved to be a daunting mission. The length and scope of the mission required full-size capital, and Tesla's particular mind and uncertainty surrounding wireless strength transmission made it tough to stable the critical financing.

Additionally, Tesla confronted technical barriers and public scepticism. The concept of wi-fi power transmission modified into as quickly as met with scepticism and scepticism from the clinical network and the everyday public. Tesla's formidable claims and his reputation as an eccentric inventor every so often hindered his capability to gather massive help and credibility. The lack of complete knowledge of his Wi-Fi energy transmission technological understanding similarly compounded the challenges he confronted. Despite the worrying conditions, Tesla pressed on with the challenge and began out creation of the Wardenclyffe Tower in Shoreham, Long Island, New York, in 1901. The tower's plan integrated a huge underground structure, massive electric powered device, and a terrific tower form. However, due to the truth the building improved, the shortage of pinnacle sufficient funding grew to become a big dilemma. Tesla's non-forestall need for added financial help strained his courting with J.P. Morgan, and the investor's enthusiasm waned as he grew to come to be more and more sceptical of the challenge's viability and profitability. The monetary difficulties, combined with Tesla's one in each of a type ventures and criminal battles, forced him to desert the Wardenclyffe Tower assignment. As the assets went into foreclosure in 1906, the tower and its gadget were sooner or later taken apart. The wi-fi strength transmission tool and global communique network that Tesla had expected in no manner got here to pass. Even even though the Wardenclyffe Tower assignment's preliminary goals weren't met, its legacy and importance stay on. The development of wi-fi communique, energy switch, and the idea of a international network had been all added on via Tesla's vision for wi-fi strength transmission, which expected technological improvements. The venture served as a testament to Tesla's unwavering dedication to pushing the boundaries of innovation within the face of big boundaries and setbacks. The Wardenclyffe net internet page has long past through tremendous efforts to keep and honour Tesla's legacy. The assets come to be offered and transformed into the Tesla Science Centre at Wardenclyffe, a museum and educational facility devoted to celebrating Tesla's contributions to technological understanding and engineering.

Although it's miles real that Tesla's improvements and achievements have been large, they weren't constantly given the credit they deserved at the identical time as he became alive. Nikola Tesla, being an enigmatic figure in data, has been surrounded with the aid of the usage of various controversies and misconceptions concerning his existence and innovations. The outstanding inventor and visionary Nikola Tesla have regularly been cloaked in rumours and falsehoods which have continued through the years. Tesla made a few superb discoveries, however it's far vital to distinguish reality from fiction and dispel advantageous myths approximately his life and legacy.

One of the most well-known controversies is the alleged competition amongst Tesla and Thomas Edison. It is frequently depicted as a battle among AC (Tesla) and DC (Edison) power systems, with Edison portrayed because the villain seeking to discredit Tesla's artwork. While there have been versions of their employer strategies and strategies, the amount of their private rivalry is exaggerated in popular subculture. Another false impression is that Tesla's Wardenclyffe Tower grow to be meant to provide the area with unfastened strength. While Tesla expected wireless power transmission, the tower itself changed into designed to illustrate the feasibility of wireless communications and to characteristic a commercial enterprise organisation task to transmit facts and indicators. It come to be no longer supposed to provide unlimited loose energy to the general public. Tesla's work on directed energy guns, regularly referred to as "lack of life rays", has caused speculation and false impression. Some receive as actual with that Tesla invented a powerful weapon capable of mass destruction. However, there can be no concrete proof to assist this declare, and Tesla's actual intentions and improvement in this kind of tool stay dubious.

Tesla's eccentric conduct and unorthodox thoughts have led to hypothesis approximately his highbrow health. Some misconceptions endorse that he changed into mentally unstable or suffered from stipulations like obsessive-compulsive contamination (OCD) or autism. While Tesla did display off notable eccentricities, there can be no conclusive evidence to useful resource the ones claims, and they frequently overlook about about his massive contributions to technological knowledge and engineering. Numerous conspiracy theories recommend that Tesla's upgrades were intentionally suppressed through effective human beings or companies because of their ability disruptive nature. These theories regularly lack credible evidence and fail to properly apprehend the complex elements that make a contribution to the success or failure of improvements. Some misconceptions characteristic the improvements and discoveries of others to Tesla by myself. While Tesla made super contributions, he worked along and constructed upon the artwork of numerous scientists and engineers. It is vital to renowned the collective efforts of the clinical community in location of attributing all achievements without a doubt to Tesla. Tesla's life is sometimes portrayed as one of unappreciated genius, with claims that he have become ignored and his enhancements stolen. While Tesla faced annoying conditions and did not normally get maintain of the attention he deserved throughout his lifetime, he did get keep of acclaim and help from severa humans, which include fantastic clients and scientists. The disputes and myths surrounding Nikola Tesla's lifestyles and achievements may be approached critically, and the ancient context should be understood. Tesla made quite some contributions, but it is important to distinguish fact from fiction and famend how collaborative generation is. We all have our own critiques and views. Therefore, separating fact from fiction is important even as examining the life and paintings of Nikola Tesla. While he have emerge as truly a first rate inventor, it's miles essential to occasionally keep away from romanticizing or sensationalizing achievements. By substantially evaluating historic information and credible belongings, we are capable of recognize Tesla's real contributions and gain a extra accurate statistics of his super legacy.

SO MUCH MORE ABOUT TESLA...

The assumption that Nikola Tesla had many untold reminiscences and facts surrounding his existence and art work is based on numerous factors. Tesla end up taken into consideration a reserved individual. He did no longer appreciably report his personal lifestyles, nor did he percentage records about his innovations and experiments. This reticence has contributed to a enjoy of intrigue and mystery surrounding his paintings. Compared to one of a kind inventors and scientists of his time, Tesla neither systematically stored meticulous records nor published his findings. Although he has authored severa patents and scientific articles, there may be a sense that hundreds of his artwork remains hidden or unexplored due to the constrained documentation available. The thriller surrounding Tesla's existence and work has been heightened thru the lack of thorough files committed to him. Over time, a number of his non-public files and lab notes had been out of region or destroyed, creating gaps in our know-how of his theories and experiments. There had been lots of rumours, myths, and speculations approximately Tesla's alleged improvements, interactions with special famous humans, and involvement with the authorities over the years. Even despite the fact that those stories regularly lack proof, they contribute to the air of mystery of thriller surrounding Tesla. Occasionally, Tesla's thoughts were considered modern or earlier of their time. Consequently, in the direction of his lifetime, some of his mind and theories were no longer really understood. Because of this, there was hypothesis about unexplored factors of his research as well as untold memories linked to his cutting-edge-day requirements.

While Nikola Tesla finished his experiments in Colorado Springs, his intense studies did result in a few unintentional outcomes. The nature of Tesla's paintings, which worried immoderate-voltage and excessive-frequency electric powered experiments, every so often pushed the boundaries of the available electrical infrastructure and had unintentional outcomes at the surroundings. Tesla used massive quantities of electricity to generate fantastically robust electromagnetic fields for his experiments at Colorado Springs. Tesla's experiments from time to time created a speedy increase in power intake that have become an excessive amount of for the nearby electricity grid, ensuing in short blackouts inside the neighbourhood. The electric electricity he have become the use of modified into so awesome that it changed into past the capability of the device on the time. One occasion, specially, modified into the said unintentional electrocution of butterflies. Tesla's high-frequency power assessments produced effective electric powered fields close to his lab. It is said that the electrical discharges electrocuted butterflies that flew too close to the take a look at. This unforeseen end result draws interest to the functionality environmental outcomes of robust electric powered fields.

It is actual that when immigrating to the us, Nikola Tesla had no everlasting house for maximum of his lifestyles. Instead, he regularly stayed in inns in particular cities, carrying out his studies and running on severa projects. Tesla has confronted monetary disturbing situations throughout his career and has struggled to maintain a sturdy income. This nomadic way of life allowed him the capability to preserve his medical pastimes with out being tied to a selected area. While can also have supplied a challenge in phrases of stability and personal comfort, it can additionally have given you the liberty to find out taken into consideration one in every of a type avenues, collaborate with precise human beings, and paintings for your innovations everywhere the need arises.

It is frequently asserted that Nikola Tesla foresaw the idea of cell phones in 1926, demonstrating his remarkable perception into wi-fi conversation. Tesla stated the opportunities of wireless era and its effects on society in an interview that changed into posted in Collier's magazine in 1926. He expected a future wherein human beings may additionally want to deliver small hand-held devices that could permit them to speak with each other wirelessly. Tesla defined a device that resembled a "pocket device" or "pocket-duration cellphone", which he believed could possibly permit humans to attach proper away, regardless of their bodily location. While Tesla's vision shared a few similarities with the concept of mobile telephones, it's miles essential to recognize that his mind were based totally completely absolutely on the thoughts of wi-fi telegraphy and his very very own experiments with the wireless transmission of alerts. The actual improvement and hobby of mobile phones as we realise them in recent times concerned the contributions of severa inventors, engineers, and technological improvements over numerous a long time. Building on in advance improvements like -way radios and vehicle phones, the idea of cell telephony, which sooner or later brought about the development of cellular telephones, commenced to take shape inside the middle of the 20th century. Cell phones, portable hand held devices with mobile network connectivity, did not grow to be commonly available till the late 20th century. It is important to distinguish between Tesla's philosophical discussions and the nice prediction of mobile phones as they exist in recent times, despite the reality that his ahead-thinking mind and preserve close to of wireless verbal exchange had been virtually in advance in their time.

The strong propensity that Nikola Tesla had for cleanliness and hygiene became an thrilling factor of his character. When it came to preserving his surroundings clean, Tesla displayed inclinations of being a germaphobe and obsessive behaviours. White gloves have been a Tesla trademark, specially whilst running with laboratory device or wearing out experiments. He concept that this manner of doing matters helped keep subjects clean and prevent contamination. Tesla had a deep-seated want for cleanliness rituals and an obsession with cleanliness. Both his hygiene and the cleanliness of his place of work had been very crucial to him. He have emerge as precise approximately the cleanliness of his lab gadget and regularly washed his palms multiple instances. Tesla disliked germs and changed into involved approximately illnesses. He took exceptional care to save you contamination and often sanitised his palms, tools, and surfaces. The large records of ailments and the price of hygiene at the time might also furthermore have contributed to his conduct as a germaphobe.

Tesla was also rumoured to dislike pearl rings because of the opportunity of infection. He notion that the tiny gaps in pearl jewellery may feature a breeding ground for microbes. Tesla saved his workspace spotless and quite organised. He made a element of stressing orderliness due to the fact he belief it promoted productivity and clarity in his artwork. His meticulous method become in direct conflict with disarray and mess. It want to also be remembered that Tesla finished X-ray experiments and considerably advanced the arena. Tesla's artwork on X-rays turned into greater state-of-the-art than previously believed, in line with present day studies. Compared to unique scientists at the time, he have become capable of produce X-rays at frequencies that have been more.

Chapter 14: The One Who Lit The World

Nikola's final days and passing were not met with the fanfare one would likely assume for this shape of first-rate mind. In New York City, Nikola Tesla died on January 7, 1943, at the age of 86. His scientific discoveries and improvements were noticeably massive, but his later years were characterized via monetary troubles and a decline in public acclaim. Tesla died in relative obscurity, however the widespread contributions he made to trendy society. During the later lengthy stretches of his existence, Tesla dwelled in extremely good New York City inns. The New Yorker Hotel, wherein he lived for ten years, become one extraordinary fame quo. Tesla stayed in inn rooms at the 33rd ground, which had been provided by manner of using the lodge. His final breath got here from this kind of rooms. The loss of life of Tesla did no longer acquire plenty of instant hobby. News of his lack of existence modified into overshadowed with the resource of the on-going occasions of World War II and media coverage have become minimum. The lack of one of the best inventors of all time end up in huge detail unknown to the overall public. Tesla's funeral modified into hung on January 12, 1943, on the Cathedral of St. John the Divine in New York City. Despite the modest attendance, his funeral changed into attended with the aid of way of leaders in technology and era that paid tribute to the character whose improvements modified the arena. Apart from this, the Tesla Company, formally called Tesla, Inc., is an American electric powered automobile and smooth electricity business enterprise based in 2003 through entrepreneurs Martin Eberhard and Marc Tarpenning. Eberhard and Tarpenning named their agency Tesla Motors, inspired via the tremendous Serbian-American inventor Nikola Tesla, who completed a critical feature inside the improvement of contemporary-day electric powered systems. The name Tesla embodied the agency's dedication to innovation and clean electricity answers.

Due to issues about feasible army makes use of of Tesla's invention, the U.S. Government took the whole thing of his property after his passing, collectively with his papers. In order to assess the significance of his examine and defend within the path of any thriller cloth moving into the wrong hands, the FBI and special corporations done inquiries. Tesla's accomplishments have obtained renewed interest and admiration within the many years for the purpose that his passing. His brilliance and in advance-thinking requirements have obtained the attention of scientists, innovators, and fans around. His legacy has been seemed with numerous museums, statues, and dedications, ensuring that his impact on society will in no manner be forgotten.

The Nikola Tesla Experience Centre has formally opened its doors inside the historic city of Karlovac in Croatia. The centre is placed near the Karlovac Grammar School constructing, which the famous inventor attended and graduated from a hundred fifty years inside the past. The centre ensures to take traffic on an immersive experience via the life and accomplishments of one of the international's great inventors, Nikola Tesla. The center combines a museum, innovation, entrepreneurship, and tourism, imparting extra younger human beings with an attractive opportunity to build up records approximately Tesla. The center covers almost 800 square meters and aspects of interactive well-knownshows, multimedia presentations, and a 3-D hologram of Tesla. The centre also has a gift maintain wherein web site website online visitors can purchase Tesla-themed souvenirs. The starting rite have become as quickly as attended by using way of way of the Prime Minister of the Republic of Croatia, Andrej Plenković, and the Minister of Culture, Nina Obuljen Korzinek. The center is anticipated to entice website traffic from across the place and can be a important visitor enchantment in Karlovac.

The centre gives current exhibitions, interactive well-known, and tasty multimedia suggests. Visitors can see Tesla's three-D hologram, it really is one of the highlights of the centre. The centre furthermore houses well-knownshows showcasing Tesla's innovative innovations, together with the Tesla coil, the rotating magnetic challenge, and the AC motor. Afterward, website online visitors can revel in multimedia shows on Tesla's visionary requirements and the impact of his artwork on cutting-edge technology. The centre additionally appears at Tesla's personal existence, which embody his love of poetry, track, and meals, in addition to his bacteriophobia and attachment to a superb pigeon. There is a present save downtown wherein traffic can buy Tesla-themed memorabilia. The centre includes a school room and an amphitheatre wherein traffic can revel in lectures and suggests about Tesla's lifestyles and artwork. Finally, the downtown café offers visitors the opportunity to reserve liquids in a futuristic manner. Well, I accept as true with that surely each person interested in Nikola Tesla's existence and artwork need to pay a visit to the Nikola Tesla Experience Centre in Karlovac. The centre's exhibits and displays offer a comprehensive and fascinating look at Tesla's floor-breaking discoveries, avant-garde thoughts, and personal existence. The facility may be a large visitor destination in Karlovac and is predicted to attract tourists from everywhere in the worldwide.

Several writers and researchers had been curious approximately Nikola Tesla's life and paintings, which has resulted within the launch of many books that find out special components of his life, innovations, and contributions to technology. These works offer assessment, biographical statistics, and in-depth analyses of Tesla's ideas and their consequences. One of the most thorough biographies of Tesla's life, "Tesla: Man out of Time" through manner of Margaret Cheney examines his early years, his career in Europe and America, in addition to his diverse inventions and experiments.

It explores Tesla's oddities, his innovative mind, and the issues he encountered in some unspecified time in the future of his career.

Similarly, "Wizard: The Life and Times of Nikola Tesla" via Marc Seifer gives an in depth examination of Tesla's existence. While "Tesla: Inventor of the Electrical Age" via using W. Bernard Carlson combines biographical details with a focal point on Tesla's improvements and their effect on modern society.

It examines Tesla's new-fangled paintings in wi-fi energy transmission, wireless communication, and alternating current (AC) structures. Additionally, it delves into the issues Tesla encountered in promoting his mind and obtaining economic backing. Whereas, "My Inventions: The Autobiography of Nikola Tesla" is a fixed of Tesla's very own writings that gives treasured insights into his thoughts, opinions, and innovations. Collectively, those works make sure that Tesla's first-rate contributions to technological information and technology can be remembered and cherished for generations to come back lower back.

www.ingramcontent.com/pod-product-compliance
Lightning Source LLC
Chambersburg PA
CBHW070734020526
44118CB00035B/1341